Lecture Notes in Economics and Mathematical Systems

Founding Editors:

M. Beckmann
H.P. Künzi

Managing Editors:

Prof. Dr. G. Fandel
Fachbereich Wirtschaftswissenschaften
Fernuniversität Hagen
Feithstr. 140/AVZ II, 58084 Hagen, Germany

Prof. Dr. W. Trockel
Institut für Mathematische Wirtschaftsforschung (IMW)
Universität Bielefeld
Universitätsstr. 25, 33615 Bielefeld, Germany

Editorial Board:

H. Dawid, D. Dimitrow, A. Gerber, C-J. Haake, C. Hofmann, T. Pfeiffer,
R. Slowiński, W.H.M. Zijm

For further volumes:
http://www.springer.com/series/300

Alexander Hübner

Retail Category Management

Decision Support Systems for Assortment,
Shelf Space, Inventory and Price Planning

Dr. Alexander Hübner
Catholic University Eichstätt-Ingolstadt
Supply Chain Management and Operations
Auf der Schanz 49
85049 Ingolstadt
Germany
alexander.huebner@ku-eichstaett.de

ISSN 0075-8442
ISBN 978-3-642-22476-8   e-ISBN 978-3-642-22477-5
DOI 10.1007/978-3-642-22477-5
Springer Heidelberg Dordrecht London New York

Library of Congress Control Number: 2011937428

© Springer-Verlag Berlin Heidelberg 2011
This work is subject to copyright. All rights are reserved, whether the whole or part of the material is concerned, specifically the rights of translation, reprinting, reuse of illustrations, recitation, broadcasting, reproduction on microfilm or in any other way, and storage in data banks. Duplication of this publication or parts thereof is permitted only under the provisions of the German Copyright Law of September 9, 1965, in its current version, and permission for use must always be obtained from Springer. Violations are liable to prosecution under the German Copyright Law.

The use of general descriptive names, registered names, trademarks, etc. in this publication does not imply, even in the absence of a specific statement, that such names are exempt from the relevant protective laws and regulations and therefore free for general use.

Cover design: eStudio Calamar S.L.

Printed on acid-free paper

Springer is part of Springer Science+Business Media (www.springer.com)

How can it be that mathematics, being after all a product of human thought independent of experience, is so admirably adapted to the objects of reality?

   Albert Einstein

Foreword

Retail shelf management forms the part of the supply chain that interfaces between the ultimate customer need and the rest of the supply chain. As such, it is regarded as the part of the supply chain where the consumer demand shows up. Matching consumer demand with supply is the core task of retailers and a key lever for increasing efficiency. Consumers demand high product availability at low prices, while retailers are ever expanding their product variety.

That is why it is not surprising that retail category management topics are gaining increasing attention – from practice and research – especially as the introduction of consumer goods launches increases year by year. The product proliferation is constrained by the limited store space and requires therefore an efficient decision making by the retailers about which products to offer and how to allocate the scare resource of shelf space. Assortment and shelf space optimization is one of the most important and difficult decisions that retailer managers have to face, as it needs to reflect consumer behavior such as substitutions, product recognition, or price sensitivity, as well as inventory, replenishment, and operational costs. Research on category management therefore intersects with research in assortment planning, inventory management, and consumer pricing. Integrating shelf space management with assortment planning and coordinating price optimization with logistics management are the core contributions of this book.

This is the first research contribution that develops shelf space management models that integrate comprehensively consumer behavior and logistical effects and analyzes under which conditions these influence the decisions and allow to improve the planning results. The book shows that not only market-related aspects but also logistical questions are impacted, if, for example, expensive shelf refill processes are required. Specifically, the dissertation develops operational methodologies for selecting optimal retail assortments, allocating it to the shelves and assigning inventory and price levels. Innovative optimization models are formulated that reflect operational constraints of shelf replenishment and are capable to solve practical relevant problem sizes.

The developed models and approaches are to be regarded as important steps toward the improvement of planning practice. I am sure that the readers will gain insights into category planning and discover a substantial addition to the emerging literature on shelf space management.

Ingolstadt *Prof. Dr. Heinrich Kuhn*

Acknowledgments

This dissertation thesis was written during my employment as external research associate in the department of Supply Chain Management and Operations at the Catholic University Eichstätt-Ingolstadt. I would like to take this opportunity to thank the numerous supporters whose backing has rendered this work possible. First of all, I am deeply committed to my advisor Prof. Dr. Heinrich Kuhn for his ongoing guidance and support. He provided fundamental input and ideas, challenged my results, and always kept me on track. I am very thankful for his excellent support and supervision as well as for our many productive discussions. His open-minded and motivating manner contributed significantly to the success of this research. I would also like to thank Prof. Dr. Joachim Büschken for serving as a co-referee of this dissertation, and for the critical revision and suggestions.

My work and life at the research group constituted an important period of my life and established many friendships. I wish to particularly thank my doctoral fellows Anna Stähr, Dr. Gerd Hahn, Christian Krudewig, Dr. Thomas Liske, Andreas Popp, Robert Schilling, Michael Sternbeck, Leopold Weckbach, Dr. Thomas Wensing, and Sven Woogt for our countless constructive discussions at research seminars and beyond. I could always count on their help. This also holds true for our team assistant Birgit Jürgens, whose appointment to our group has proven to be a great enrichment, not only because we grew up in the same region.

A special thanks goes to the German Ministry of Education and Research, the German Academic Exchange Service, the Hanns-Seidel-Foundation and McKinsey & Company, who have supported me with doctoral scholarships during the dissertation.

As a child I never thought I will be engaged with retailing, as my parents did their entire business career. But the world is full of surprises. Many thanks to their support and inspiration, and that I got in touch with shelf space quite early.

I would like to thank especially my partner Caroline Graßold for her love, understanding, and continuous encouragement, which I always could rely on. I am looking forward to our new ways and joint objectives in our life.

Ingolstadt *Alexander Hübner*

Executive Summary

Retail shelf management means cost-efficiently matching retail operations with consumer demand. As consumers expect high product availability and low prices, and retailers are constantly increasing product variety and striving towards high service levels, the complexity of managing retail business and its operations is rocketing. Consumers demand literally meets the retailers offering at the point of sales – at the shelf. Retailers need to match consumer demand with shelf supply by balancing variety (number of products), service levels (number of items of a product), and optimizing demand and profit via carefully calibrated prices. As a result the core strategic decisions a retailer must take involve assortment sizes *(listing)*, shelf space management *(facing)*, *replenishing* and *pricing*. For example, offering broader assortments may limit the appropriate service levels and vice versa. Lower item prices result in a lower profit contribution per item sold, but increase demand, which needs to be fulfilled appropriately. Common practice in retail is to make decisions about the planning problems sequentially: Retailers first determine the assortment size, next allocate it to the shelves, assign prices, and then finally make arrangements for instore shelf replenishment. However, as the problems are interrelated, managing these operations in isolation may result in suboptimal decisions if communication flows are omitted. The focus is consequently on integrated shelf space management models reflecting these interrelated issues.

The contribution of this dissertation is to develop rigor and comprehensive decision models for shelf space management and its subproblems of listing, facing and pricing of products. The profit maximization models address the assortment, space allocation, replenishment and pricing decisions for a set of products. Assortment planning deals with the *listing* decision to determine which products should be included in the assortment based on substitution effects. Shelf space management addresses the space assignment for individual products *(facing)* based on space-elasticity effects and restocking frequencies and costs *(replenishing)*. Finally, *pricing* utilizes price-elasticity effects to steer consumer demand. Key assumptions of the models include retailer perspective, mid-term planning horizons, deterministic demand and efficient replenishment systems to avoid stock outs.

A comprehensive framework for retail demand and supply chain planning is provided to put these models into a broader context. The planning matrix structures and identifies long-term to short-term planning problems in the domains of procurement, warehousing, distribution and sales. The recent past has witnessed exciting new research – both theoretical and applied – aimed at addressing retail shelf space management problems. This dissertation therefore also summarizes state-of-the-art empirical insights, quantitative models and software applications for shelf space management.

Contents

1 Outline 1
 1.1 Background and Motivation 1
 1.2 Objectives of this Research 3
 1.3 Classification and Contribution.. 3
  1.3.1 Scope 3
  1.3.2 Methods Applied.. 4
 1.4 Overview of the Chapters.. 5
  1.4.1 Framework for Retail Demand and Supply
   Chain Planning.. 6
  1.4.2 Empirical Insights, Quantitative Models
   and Software Applications for Master Category Planning.... 7
  1.4.3 Shelf Space and Assortment Planning 8
  1.4.4 Shelf Space, Assortment and Inventory Planning 10
  1.4.5 Shelf Space, Assortment and Price Planning 11
  1.4.6 Conclusions and Outlook...................................... 12

2 Framework for Retail Demand and Supply Chain Planning 15
 2.1 Introduction 15
 2.2 Contribution to Planning Frameworks 16
  2.2.1 Research Objective 16
  2.2.2 Integral Planning Perspective at Entire
   Retail Supply Chain 18
  2.2.3 Comprehensive Quantitative Decision Support Systems 18
 2.3 Specifics of Grocery Retailing... 19
  2.3.1 Attributes of the Retail Grocery Supply Chain 19
  2.3.2 Reasons for Modifications in Retail Planning 20

xiii

	2.4	Framework for Retail Demand and Supply Chain Planning	22
		2.4.1 Overview of Retail Demand and Supply Chain Planning	22
		2.4.2 Long-Term Configuration Planning	23
		2.4.3 Mid-Term Master Planning	26
		2.4.4 Short-Term Execution Planning	32
		2.4.5 Summary of the Retail Demand and Supply Chain Planning Framework	35
	2.5	Aspects of Planning Interdependencies at Retail Shelf Management	38
		2.5.1 Example: Vertically Integrated Shelf Space and Price Management	38
		2.5.2 Example: Horizontally Integrated Retail Operations Form Warehouse to Shelf	39
	2.6	Conclusions and Future Areas for Research	39
		2.6.1 General Application of Retail DSCP Matrix	39
		2.6.2 Unified Modeling Structure	40
		2.6.3 DSSs for Dedicated Planning Problems	40
		2.6.4 DSSs for Interrelated Planning Problems	41
		2.6.5 Implementation of Advanced Models in Commercial Software Packages	41
3	**Empirical Insights, Quantitative Models and Software Applications for Master Category Planning**		**43**
	3.1	Introduction to Master Category Planning	43
	3.2	Definition and Scope of Master Category Planning	45
	3.3	Software Applications for Master Category Planning	45
		3.3.1 Popularity of Software Systems in Category Planning	45
		3.3.2 Scope and Overview	47
		3.3.3 Limitation of Commercial Software Applications	49
	3.4	Scientific Models for Master Category Planning	50
		3.4.1 Assortment Planning Models	51
		3.4.2 Shelf Space Planning Models	60
	3.5	Conclusions and Future Areas for Research	65
		3.5.1 Alignment of Software Applications and Science	67
		3.5.2 Alignment of Assortment and Shelf Space Management	68
		3.5.3 Alignment with Other Planning Objectives	68
		3.5.4 Alignment within Shelf Space Competition	68
		3.5.5 Summary	69
4	**Assortment and Shelf Space Planning**		**71**
	4.1	Introduction and Motivation	71
	4.2	Problem Definition	72
		4.2.1 Properties of Demand Function	73
		4.2.2 Instore Inventory Management and Shelf Replenishment	75

	4.3	Literature Review of Assortment and Shelf Space Planning	77
		4.3.1 Assortment Planning Models	77
		4.3.2 Shelf Space Planning Models	78
	4.4	Formulation of the Capacitated Assortment and Shelf Space Problem (CASP)	79
		4.4.1 Objective Function	79
		4.4.2 Constraints	81
	4.5	Numerical Examples and Test Problems	82
		4.5.1 Illustrative Example for Impact of Substitution	82
		4.5.2 Test Case for Hard Knapsack Problem	83
		4.5.3 Applicability of CASP for Large-Scale Problems	89
	4.6	Conclusions and Future Areas for Research	90
5	**Assortment, Shelf Space and Inventory Planning**		93
	5.1	Introduction and Motivation	93
	5.2	Problem Definition	95
		5.2.1 Inventory Management Systems and Cost Implications	95
		5.2.2 Properties of Demand Function	97
	5.3	Literature Review of Shelf Space and Inventory Planning	97
	5.4	Formulation of the Capacitated Assortment, Shelf Space and Replenishment Problem (CASRP)	100
		5.4.1 Objective Function	100
		5.4.2 Constraints	102
	5.5	Numerical Examples and Test Problems	103
		5.5.1 Description of the Test Case	103
		5.5.2 Profit Impact of Integrated Assortment, Shelf Space and Inventory Planning	104
		5.5.3 Impact on Solution Structure	105
		5.5.4 Sensitivity Analyses of CASRP	106
		5.5.5 Applicability of CASRP for Large-Scale Categories	111
	5.6	Conclusions and Future Areas for Research	112
6	**Assortment, Shelf Space and Price Planning**		113
	6.1	Introduction and Motivation	113
	6.2	Problem Definition	114
		6.2.1 Properties of Demand Function	116
		6.2.2 Instore Inventory Management and Shelf Replenishment	120
	6.3	Literature Review of Shelf Space and Price Planning	120
	6.4	Formulation of the Capacitated Assortment, Shelf Space and Price Problem (CASPP)	123
		6.4.1 Objective Function	123
		6.4.2 Constraints	125

	6.5	Numerical Examples and Test Problems	127
		6.5.1 Description of the Test Case	127
		6.5.2 Profit Impact of Integrated Assortment, Shelf Space and Price Planning	128
		6.5.3 Further Managerial Insights	129
		6.5.4 Sensitivity Analyses	130
		6.5.5 Applicability for Large-Scale Categories	131
	6.6	Conclusions and Future Areas for Research	132
7	**Conclusions and Outlook**		**135**
	7.1	Contribution to Research Status	135
	7.2	Further Areas for Research	137
		7.2.1 Alignment with Other Planning Problems	137
		7.2.2 Further Demand Effects	138
		7.2.3 Empirical Validation of Model Recommendations	139
		7.2.4 Modeling Techniques	139
		7.2.5 Transfer to Commercial Software Applications and Retail Practice	140
Bibliography			143

Notation

Indices:

$i = 1, \ldots, I$	Index of items, with $i \in N^+$ ($i \in N^-$) as the listed (delisted) items
$k = 1, \ldots, m, \ldots, K$	Index of facing levels; k equals the number of facings m is the index of the base-facing level, $k = m$, i.e., the number of facings observed
$l = 1, \ldots, n, \ldots, L$	Index of price levels; n is the index of the base price level, $l = n$, i.e., price level observed

Parameter:

α_i	Base demand of item i
β_i	Space elasticity of item i
δ_{ikl}	Substitution weight of item i depending on facing level k and price level l
ϵ_{il}	Price elasticity of item i at price level l
γ_{ji}	Cross-space elasticity between item j and item i
λ_j	Percentage of demand which is latently if item j is delisted
μ_{ji}	Substitution rate between item j and i
η_j	Fraction of consumers who are not willing to compromise their initial choice for product j
a_i	Probability that item i is available at customer arrival
BSL	Basic supply level achieved by regular scheduled shelf filling
b_i	Breadth of item i
C	Customer with $C = 1, 2, \ldots, C^{max}$
$\overline{C}$	Mean number of customers visiting store
c_i	Unit costs of item i
DG_i (DL_i)	Demand gain (demand loss) of an item i through changes of the merchandizing variables of item j
d_i	Demand of item i

d_i^{max}	Maximum demand of item i
d_{ik}	Demand of item i at facing level k
d_{ikl}	Demand of item i at facing level k and price level l
$\tilde{d}_{j0}$	Latent demand if item j is delisted
FCR_i	Fixed costs of replenishment of item i
FOC_i	Fixed order costs of item i
f_i	Probability that an arrivaing customer will initially prefer item i
g_i	Number of units of item i supplied through basic refill process per facing
h	Interest rate for inventory holding costs
$k_i^{max}(k_i^{min})$	Upper (lower) bound on the number of facings of item i
LC_i	Fixed listing costs of item i
MS_i	Marketshare of item i
N	Item set with $N = \{1, 2, \ldots, i, \ldots, I\}$, with N^+ (N^-) as the set of listed (delisted) items
p_i	Unit profit (gross margin) of item i
p_{il}	Unit profit (gross margin) of item i at price level l; $p_{il} = r_{il} - c_i$
q_i	Stocking quantities of item i
$q_i^{(o)}$	Overstocked inventory of item i
$q_i^{(u)}$	Undersupplied inventory of item i
RF	Number of basic refill frequency, e.g., shelf restocks in morning
RFC_i	(Variable) Refill costs of item i
r_i	Unit price of item i
$r_{il}(r_{in})$	Unit (base) price of item i at price level l (at base price level n)
$r_i^{max}(r_i^{min})$	Upper (lower) bound on the prices of item i
S	Available shelf space
T	Period
u_i	Consumer utility of item i
v_i	Consumer preference of item i
v_{ik}	Minimum sales volume per item i for each facing k

Decision variables:

y_{ik}	Binary variable of item i at facing level k, which is set to value 1, if item-facing combination is chosen, otherwise 0
y_{ikl}	Binary variable of item i at facing level k and price level l, which is set to value 1 if an item-facing-price combination is chosen, otherwise 0
z_i	Binary variable of item i, which is set to value 1, if item is listed, otherwise 0

Abbreviations

BM	Base model
BSL	Basic supply level
CASP	Capacitated assortment and shelf space problem
CASPP	Capacitated assortment, shelf space and price problem
CASRP	Capacitated assortment, shelf space and replenishment problem
CSP	Capacitated shelf space problem
CM	Category management
DSCP	Demand and supply chain planning
DSS	Decision support system
ECR	Efficient consumer response
GA	Genetic algorithm
GRG	Generalized reduced gradient algorithm
IIA	Independence of irrelevant alternatives
LP	Linear problem
MINLP	Mixed-integer non-linear problem
MIQP	Mixed-integer quadratic problem
MIP	Mixed-integer problem
MNL	Multinominal logit
OOA	Out-of-assortment
OOS	Out-of-stock
SC	Supply chain
SCM	Supply chain management
SCP	Supply chain planning
TP	Total profit
TCL	Total costs of listing
TCOI	Total costs of overstocked inventory
TCPP	Total cross-product profit
TCSP	Total cross-space profit
TCUS	Total costs of undersupply
TDP	Total direct profit
TSP	Total substitution profit

List of Figures

Fig. 1.1	Overview of chapters	5
Fig. 1.2	Retail demand and supply chain planning framework – overview	6
Fig. 2.1	Business-to-Business supply chain planning framework (Fleischmann et al. 2008)	19
Fig. 2.2	Retail demand and supply chain planning framework – overview	22
Fig. 2.3	Retail supply chain network	23
Fig. 2.4	Retail demand and supply chain planning framework – detailed view	36
Fig. 2.5	Retail demand and supply chain planning framework – summary	37
Fig. 3.1	Interdependencies in master category planning	46
Fig. 3.2	Retailers management capabilities in merchandising (2009)	47
Fig. 3.3	Retail category planning applications – vendors' functionalities and strengths	48
Fig. 3.4	Summary: State-of-the-art and future areas for research	67
Fig. 4.1	Example for estimate of latent demand	75
Fig. 4.2	Facing-dependent supply and demand curve without substitution and cross-space effects	76
Fig. 4.3	Illustrative example for integrated assortment and shelf space planning	83
Fig. 4.4	Profit-space correlation for assortment and shelf space test problem	85
Fig. 4.5	Impact of integrated assortment and shelf space planning	87
Fig. 4.6	Impact of BSL constraint	88

Fig. 4.7	Impact of cross-space effects: $CASP_{CR}$ vs. CASP	88
Fig. 4.8	Sensitivity analyses of the $CASP_{CR}$-model	89
Fig. 5.1	Comparison of space-dependent demand and supply	96
Fig. 5.2	Development of retail shelf inventory levels	97
Fig. 5.3	Profit impact of integrated assortment, shelf space and inventory planning	105
Fig. 5.4	Impact of inventory-related costs on solution structure	106
Fig. 5.5	Items with reduced facings in CASRP model	107
Fig. 5.6	Additionally listed items in CASRP model	107
Fig. 5.7	Sensitivity analyses of the CASRP for managerial planning aspects	108
Fig. 5.8	Sensitivity analyses of the CASRP for cost parameters	109
Fig. 5.9	Sensitivity analyses of the CASRP for consumer behavior	110
Fig. 6.1	Decision ownership and interaction with competition	115
Fig. 6.2	Example: Cross-product demand shifts	119
Fig. 6.3	Comparison of facing and pricing values in the CASPP and CASP models	129
Fig. 6.4	Improvement of shelf utilization through the CASPP-model	130
Fig. 6.5	Sensitivity analyses of the CASPP-model	131

List of Tables

Table 1.1	Shelf space demand effects and supply considerations	4
Table 1.2	Core criteria of literature streams in assortment and shelf space planning	8
Table 1.3	Integrated effects in models	9
Table 3.1	Overview of major CM software based on Griswold (2007) and own analyses	48
Table 3.2	Core criteria of literature streams in assortment and shelf space planning	50
Table 3.3	Overview of empirical studies on substitution behavior	52
Table 3.4	Overview of literature on assortment models	57
Table 3.5	Overview of literature on shelf space management models	66
Table 4.1	Demand types of the CASP-model	73
Table 4.2	Substitution matrix of illustrative example	83
Table 4.3	Data for assortment and shelf space test problem	84
Table 4.4	Profit impact and run time of integrated models with varying items and facings	86
Table 4.5	Evaluation of computation performance of the CASP-model	90
Table 5.1	Retail shelf space management models related to inventory management	100
Table 5.2	Data for assortment, shelf space and replenishment test problem	104
Table 5.3	Evaluation of computation performance of the CASRP-model	111
Table 6.1	Sales and profit data for assortment, shelf space and price test problem	128

Table 6.2	Profit impact of integrated assortment, shelf space and price planning	129
Table 6.3	Evaluation of computation performance of the CASPP-model	132

Chapter 1
Outline

The objective of this thesis is to analyze and develop decision support systems for retail shelf space management. To provide the context, a comprehensive retail operations planning framework is developed for retail demand and supply chain planning. A review of empirical insights, quantitative decision support systems and commercial software applications indicate the need for more advanced models reflecting category managers' actual decision problems. The focus of this research is therefore to examine retail shelf space problems and develop decision support systems, that support assortment planning (*which products to offer?*), shelf space planning (*how much space to allocate to products?*), inventory planning (*how to align restocking requirements?*) and price planning (*which price to assign to each product?*) to maximize the profitability of a retail grocery category.

This introductory chapter specifies the background and motivation (1.1), derives the research objectives (1.2) and contribution to the state of research (1.3), followed by an overview of the individual chapters (1.4).

1.1 Background and Motivation

Matching consumer demand with retail shelf supply is a key lever for increasing efficiency in the retail industry. Consumers are demanding better service levels and prices, while retailers respond with increasing product variety, becoming more price competitive and striving towards higher service levels. These developments have greatly increased pressure on margins and the complexity of managing retail shelf space, which may be one of the scarcest and most strategically valuable operational resources. The category manager's objective is how best to organize product assortments and merchandizing plans to generate greater profit contribution from their existing, limited shelf space, especially as the increasing product variety is in conflict with limited shelf space.

In addition to its classical supply function, shelf inventory has a demand-generating function, as more facings lead to increasing consumer demand through higher product visibility (e.g., Inman et al. 2009, Chandon et al. 2009). Increasing the number of facings for one product limits the space available to other products, and may require the delisting of other products. This means also that latent consumer demand of the delisted products cannot be directly satisfied, and consumers may switch to other products or outlets. The *demand side* and the *supply side* therefore need to be aligned. For example, marketing activities such as price adjustments increase or lower demand, and overall supply and product availability can be influenced by adjusting replenishment frequency.

Empirical studies demonstrate the benefit of comprehensive shelf space management. McIntyre and Miller (1999), Hennessy (2001), Basuroy et al. (2001), Dhar et al. (2001), Shugan and Desiraju (2001), ECR Europe (2003a), Levy et al. (2004), Grocery Manufacturers Association et al. (2005), Desrochers and Nelson (2006), Campillo-Lundbeck (2009) and Fisher and Raman (2010) empirically show that well-executed shelf space management has positive profit implications in terms of an efficient consumer response, and the subsequent emphasis on category management. Several trends have made the issue of an efficient shelf management system to one of the most critical marketing and operational decisions (Levy et al. 2004; Hall et al. 2010). German retailers and consumer goods producers recently rated the "optimization of product portfolio and category management" as the most important task for achieving performance goals according to a survey of McKinsey & Company (Breuer et al. 2009). This is not surprising as shelf space competition in retail stores is at an all-time high, driven by the competitive need to constantly introduce new products. Since the 1990s, there has been significant product proliferation. The average number of items in overall store assortments increased by 20% between 1970 and 1980, and by 75% between 1980 and 1990 (Greenhouse July 17, 2005). In confectionery, for example, the number of brands rose by more than 40% between 1997 and 2001, but overall volume sales by only 0.8% (Carlotti et al. 2006). Additionally, most retailers suffer from decreasing space productivity. Gutgeld et al. (2009) concluded that 19 out of 24 European retailers were unable to maintain their space productivity. Retailers are under significant pressure to improve their operational efficiency and offer competitive prices. Furthermore, as shoppers increasingly take their purchase decisions instore, retail marketers are diverting a growing proportion of their marketing budgets from traditional out-of-store media advertising to instore marketing (Xin et al. 2009; Chandon et al. 2009). Consequently, it is becoming extremely important for retailers how they manage their shelf space.

A key imperative to achieving profitable shelf space arrangements also depends on category managers' access to efficient decision support systems to manage their shelf space. The traditional shelf space management tool employed by retailers is a planogram. Software applications generate without time-consuming computations shelf layout recommendations based on simplistic "rules of thumb" like allocating space proportional to sales. However, as Lim et al. (2004) point out, "due to the problem's complexity, only relatively simple heuristic rules have been developed and are available for retailers to plan product-to-shelf allocation (Zufryden 1986; Yang 2001) (...) These are not effective global optimization tools (Desmet and Renaudin 1998) and are largely used for planogram accounting to reduce time spent on manual manipulation of shelves (Drèze et al. 1994; Yang 2001)" (see also Irion et al. 2004; Griswold 2007; Kök et al. 2009; Murray et al. 2010; Hansen et al. 2010). Their main purpose is the simulation of alternative item placements on-screen and the related profit analyses. Kopalle (2010) notes that "substantial" sales could be lost by retailers who rely on such simple heuristic rule-based decision support systems.

1.2 Objectives of this Research

The objective of this thesis is to develop effective and efficient decision support systems for retail shelf space management. An efficient model needs to reflect merchandise variables that impact consumer demand and logistical components of shelf replenishment of fast-moving consumer goods. Fast-moving consumer goods are products that are sold quickly, replenished regularly and sold at relatively low cost with higher price sensitivity.

In demand and supply chain management, analytical methods are emerging as promising solutions to many of these planning problems (Agrawal and Smith 2009b; Fisher and Raman 2010). An analytics orientation at many retail organizations provides a great opportunity for modelers and is nudging retail managers towards more quantitative decision making (Kopalle 2010). However, retail shelf space practice and research lack a holistic planning architecture that is based on an *integral planning view* on retail requirements and constraints (i.e., structuring planning tasks hierarchically and taking into account operational constraints), *integrated consumer instore behavior* (i.e., integrating decision-relevant consumer behavior based on empirical insights into planning problems), and *comprehensive quantitative decision support* (i.e., using decision support systems based on appropriate quantitative models and proper optimization). Therefore, the research questions targeted in this work are:

1. How should a comprehensive retail demand and supply chain planning framework be configured that integrates horizontal and vertical planning interdependencies?
2. What are state-of-the art empirical insights, quantitative decision support systems and commercial software applications for shelf space management?
3. How can efficient shelf space management models be developed that reflect category managers decision problems?

1.3 Classification and Contribution

1.3.1 Scope

This dissertation is – to the best of the author's knowledge – the first coherent contribution that structures retail shelf space management problems from multiple perspectives and develops comprehensive quantitative decision support models. The contribution of this research is therefore also to structure the planning problems and provide appropriate solution methods.

Table 1.1 Shelf space demand effects and supply considerations

Decision problem	Decision area	Effects
Assortment planning	Assortment size and listing of products	Substitution demand
Shelf space planning	Shelf space allocation and number of facings	Space-elastic demand
Inventory planning	Restocking frequency	Supply constraints and costs
Price planning	Pricing of products	Price-elastic demand

The category manager's shelf space decision problem can be characterized as a multi-perspective same-time decision problem, where a number of questions need to be solved jointly: what to list (assortment planning), how much of the products to put into the shelf (shelf space planning), how often to replenish (inventory planning), and how other marketing effects such as price can impact demand (price planning). This thesis will introduce models that integrate some of the most relevant instore consumer choice and instore logistic effects:

This research contributes to the body of shelf space management literature by offering a more comprehensive demand and cost analyses. Current publications solve the interdependent decision by focusing mainly on isolated issues or abstracting from the category managers actual problems. Shelf space management literature deals mainly with space effects for additional space, but does not integrate substitution. One issue is that, the literature on assortment deals mainly with substitution effects in the case of non-available items, without considering space effects. Another is that, shelf space management models assume that replenishment systems are efficient, and do not apply decision-relevant restocking costs. However, shelf space and restocking planning are interdependent, e.g., low space allocation requires frequent restocking. Furthermore, demand effects of price adjustments have not been studied in shelf space management models (Table 1.1).

Despite retail managers striving to follow the mantra "retail is detail," most retail managers have little time to consider the details of different category arrangements. Consequently, the benefits of using shelf space models to supplement human decision making depend on how efficiently and quickly the shelf space models can run on computers. Researchers have reduced the necessary solution time by applying specialized heuristic or meta-heuristic search algorithms to the basic shelf space allocation parameters (Hansen et al. 2010). These methods may be appropriate as the shelf space allocation problem is in many cases "NP-hard" (i.e., nondeterministic polynomial-time hard). However, these algorithms do not guarantee globally optimal objective values.

1.3.2 Methods Applied

We study multi-product mid-term and deterministic shelf space, assortment, price and inventory management problems that integrate facing-dependent demand effects, substitution effects, price effects, as well as inventory holding and replenishment costs. The shelf space management models are studied as mixed-integer non-linear problems.

We transfer the problems into mixed-integer knapsack problems, which then allow the efficient use of standard solvers such as CPLEX. The models are tested with empirical consumer and retailer data to investigate the viability of the model, provide practical decision support systems, and demonstrate their superiority over commonly used industry applications. The numerical examples illustrate the benefit of an integrated decision model and pave the way for decision making for typical retail categories, and the implementation of such a model in commercial software applications.

1.4 Overview of the Chapters

This dissertation is divided into six more chapters: Chap. 2 provides a comprehensive retail operations planning framework to set the context for demand and supply chain planning. The next Chap. 3 reviews empirical insights, academic quantitative decision support systems, and commercial software applications for shelf space management. Chapters 4–6 are concerned with developing shelf space management models. The content of these chapters is briefly presented in Fig. 1.1:

Chapter 2	Chapter 3	Chapter 4	Chapter 5	Chapter 6	Chapter 7
Planning Framework for Retail Demand and Supply Chain Planning	Empirical Insights, Quantitative Models and Software Applications for Master Category Planning	Assortment and Shelf Space Planning	Assortment, Shelf Space and Inventory Planning	Assortment, Shelf Space and Price Planning	Conclusions and Outlook
Structure					
2.1 Introduction 2.2 Overview of Planning Frameworks 2.3 Specifics of Grocery Retailing 2.4 Framework for Retail Demand and Supply Chain Planning 2.5 Aspects of Planning Inter-dependencies 2.6 Conclusions and Future Areas for Research	3.1 Motivation 3.2 Definitions and Scope of Master Category Planning 3.3 Software Applications for Master Category Planning 3.4 Scientific Models for Master Category Planning 3.5 Conclusions and Future Areas for Research	4.1 Introduction 4.2 Problem Definition 4.3 Literature Review 4.4 Model Formulation 4.5 Numerical Examples and Test Problems 4.6 Conclusions and Future Areas for Research	5.1 Introduction 5.2 Problem Definition 5.3 Literature Review 5.4 Model Formulation 5.5 Numerical Examples and Test Problems 5.6 Conclusions and Future Areas for Research	6.1 Introduction 6.2 Problem Definition 6.3 Literature Review 6.4 Model Formulation 6.5 Numerical Examples and Test Problems 6.6 Conclusions and Future Areas for Research	7.1 Contribution to Research Status 7.2 Future Areas for Research
Objectives					
Comprehensive architecture for retail demand and supply chain planning	Review of empirical studies, quantitative decision support systems and software	Development of an integrated model for shelf space and assortment planning	Development of an integrated model that also reflects supply requirements	Development of an integrated model that also reflects price-demand adjustments	Summary of research results and future research opportunities

Fig. 1.1 Overview of chapters

… Framework for Retail Demand and Supply Chain Planning

…nd chapter provides the context for retail shelf space planning: It develops a comprehensive operations planning framework, identifies interdependencies, and structures demand and supply planning problems. First results have been also published in the working paper Hübner et al. (2010).

Retail demand and supply chain planning means matching cost-efficient consumer demand with retail operations. As consumers are demanding better service levels and purchase prices, and retailers are increasing their channel formats, product variants and taking over a growing number of logistics operations from producers, the complexity of managing retail business and its operations is spiraling.

The contribution of this chapter is therefore to structure retail demand and supply chain planning tasks holistically from the point of consumption to the point of production in order to develop new insights for management practice and retail research. Retail practice and research lack a holistic architecture based on an integral planning view on the entire retail supply chain. It needs to structure planning tasks hierarchically along the supply chain and taking into account organizational interdependencies. Secondly, it also means using comprehensive decision support systems based on appropriate quantitative models and proper optimization.

This chapter expands supply chain planning with retail specifics such as consumer interaction and warehousing. A corresponding consumer-backed demand and supply chain planning architecture is developed to structure the interrelated planning issues. Planning modules are derived in the domains of procurement, warehousing, distribution and sales, for long-term configuration, mid-term master planning and short-term fulfillment. The architecture consists of dedicated interrelated planning modules that decentralize decision making for complexity-related and organizational reasons. Information flows along planning modules are essential, and the value of comprehensive quantitative planning methods will be demonstrated.

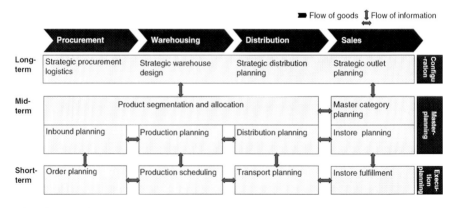

Fig. 1.2 Retail demand and supply chain planning framework – overview

The analyses show the need to develop retail-tailored analytical models. This will improve planning quality and generate efficiency gains for the entire business. The study indicates the need for decision models with integrated hierarchical aspects (e.g., mid-term category management and operative instore fulfillment), sequential planning aspects (e.g., impact of warehouse picking on instore logistics), and the implementation of these in commercial software packages (Fig. 1.2).

The following chapter therefore provides a more targeted review of empirical studies, quantitative models and software applications in the field of master category planning.

1.4.2 Empirical Insights, Quantitative Models and Software Applications for Master Category Planning

Chapter 3 introduces commercial software for shelf space management and reviews scientific models of category management, mainly including retail assortment planning and retail shelf-space allocation. The primary retailer's objective in category management is to profitably increase and satisfy consumer demand. Retailers and researchers are increasingly addressing master category planning with quantitative models (Kopalle 2010). Empirical studies illustrate the importance of optimizing the assortment, shelf space allocation and replenishment (e.g, Drèze et al. 1994; Boatwright and Nunes 2001; Gruen et al. 2002; Ketzenberg et al. 2002; Sloot et al. 2005; Koschat 2008; Xin et al. 2009; Chandon et al. 2009).

The rapid development of advanced scientific models and software applications, particularly for space management, has created a wide range of decision support systems in this field. Retailers and producers are using software applications to create assortment and merchandising plans. The popularity of these applications is mainly due to their simplicity in implementing decisions for a large number of items and visualizing shelf arrangements. The shelf recommendation is mainly based on a proportional space allocation rule. A manual and iterative search is then conducted by repeatedly creating planograms, which does not guarantee that the optimal solution will be found.

The goal of this chapter is therefore to identify, structure and examine the decision support systems for assortment planning and shelf space management. Consistent with previous research, the two areas are defined as follows:

- *Assortment planning*: Listing decisions based on consumer choice behavior and substitution effects.
- *Shelf space planning*: Facing and replenishment decisions based on space- and cross-space elasticity effects, limited shelf space and operational restocking constraints.

These planning questions are inevitably interdependent. For example, large assortments drive lower inventory levels of individual items, which reduce the visibility of those items on the shelves, increase the risk of stock-outs and impose

Table 1.2 Core criteria of literature streams in assortment and shelf space planning

Criteria	Assortment planning	Shelf space planning
Substitution effects	X	(X)
Space-elastic demand		X
Limited shelf space	(X)	X

X: fully integrated; (X): partially integrated

high restocking costs due to the need for frequent replenishment (Campo and Gijsbrechts 2005; Mantrala et al. 2009). It will be shown that these planning problems are not yet sufficiently and comprehensively integrated in commercial software systems and scientific models.

This chapter analyzes different properties of the various streams in assortment and shelf space planning (see Table 1.2). We show that retail assortment planning models neglect to integrate space-elastic demand and partially the constraints of limited shelf space. Shelf space management streams also omit substitution effects between items, when products are delisted or out of stock, which is the focus of consumer choice models in assortment planning.

The challenge for retailers and researchers lies in taking an integrated view when developing demand-and supply-oriented planning models. There are difficulties commonly involved in the use of commercial software and the implementation and transfer of scientific models. Addressing these issues, this chapter provides a state-of-the-art overview and research framework for integrated assortment and shelf space planning. A more rigorous quantitative approach to merchandising will allow the tight alignment of retail shelf supply with consumer demand, leading to efficiency gains and more productive shelf space (see also Hübner and Kuhn 2012).

The following three chapters develop shelf space models that contribute to more comprehensive modeling.

1.4.3 Shelf Space and Assortment Planning

Analyses of empirical consumer behavior studies as well as the scientific decision support systems and software applications used up to now indicate the need for more practical solution methods for shelf space and assortment planning. Chapter 4 develops a model for integrated shelf space and assortment planning. This constitutes a basic model for the subsequent chapters, which is also partially published in Hübner and Kuhn (2011d).

Additional models follow the emergence of further extensions and the integration of additional effects and constraints. Chapter 5 extends the basic model by *supply-side* considerations and integrating restocking costs, while Chap. 6 extends the model by *demand-side* considerations and integrates price optimization into shelf space management (Table 1.3).

1.4 Framework for Retail Demand and Supply Chain Planning

Table 1.3 Integrated effects in models

Criteria	Chap. 4	Chap. 5	Chap. 6
Substitution demand	X	X	X
Space-elastic demand	X	X	X
Limited shelf space	X	X	X
Restocking costs		X	
Price-elastic demand			X

Efficiently managing limited retail shelf space is critical as the increase in product variety is in conflict with limited shelf space and operational costs. Chapter 4 presents a multi-product shelf space management problem, where category demand is a composite function of shelf space allocated and consumer-driven demand substitution under replenishment constraints. It describes an innovative approach to solving a core retail planning question involving which products to offer and how much shelf space to assign items. The objective function is formulated as follows, with P as the category profit, consisting of *total direct profit* (TDP), *total substitution profit* (TSP) and *total costs of listing* (TCL).

$$\text{Max! } P(\bar{k},\bar{z}) = \text{TDP} + \text{TSP} - \text{TCL} \tag{1.1}$$

The TDP function in (1.2) reflects space-elastic demand d_i, depending on the number of facings k_i allocated to item i and the item profit p_i. The demand rate of item i is a deterministic function of its displayed front row inventory level. The basic demand is denoted by α_i. In accordance with prior research, the space-demand relationship β_i is modeled with elasticity as a power function. b_i is the breadth of an item i. The TSP function in (1.3) reflects substitution demand $d_j(z_j = 0)$ from a delisted item j to item i, expressed by the latent demand of the item j and the substitution probability μ_{ji} between the items, and with z_i as a binary variable set to 1 when items are included in the assortment. The TCL function in (1.4) applies listing costs LC_i for all listed items.

$$\text{TDP}(k_i) = \sum_{i}^{I} d_i(k_i) \cdot p_i \quad \text{with} \quad d_i(k_i) = \alpha_i \cdot (k_i \cdot b_i)^{\beta_i} \tag{1.2}$$

$$\text{TSP}(\bar{k},\bar{z}) = \sum_{i}^{I} \sum_{\substack{j=1 \\ j \neq i}}^{I} d_j(z_j = 0) \cdot \mu_{ji} \cdot p_i \cdot z_i \tag{1.3}$$

$$\text{TCL}(z_i) = \sum_{i=1}^{I} z_i \cdot \text{LC}_i \tag{1.4}$$

Major constraints are in the number of facings, limited total shelf space and the basic supply level (BSL). The latter defines minimum supply levels, which need to be achieved via a scheduled refilling before the sales period. Only a limited share (1-BSL) of the demand can be refilled additionally during a sales period.

$$\text{Supply}_i(k_i) = \text{Demand}_i(\bar{k}) \cdot \text{BSL} \qquad i = 1, 2, \ldots, I \qquad (1.5)$$

Traditional shelf space models are extended in three directions. First, the model takes into account facing-dependent restocking constraints. Second, the model integrates substitution effects by modeling out-of-assortment substitution and introducing a zero-space demand estimate. The mixed-integer non-linear problem is then transformed into a knapsack structure. The model can be used to solve hard knapsack problems and large problem instances of theoretically entire stores. The computation tests show that integrating assortment effects into shelf space management avoids wasting space and frequent restocking situations. This integrated approach jointly improves product availability and profit, and provides more accurate results for the underlying consumer behavior. Hence it provides a practical solution for category-specific problem sizes.

The model is extended by investigating further shelf-*supply* aspects and the impact of inventory management on shelf space management in Chap. 5, while further shelf-*demand* aspects and the impact of price management on shelf space management are examined in Chap. 6.

1.4.4 Shelf Space, Assortment and Inventory Planning

Current models for shelf space management assume unlimited shelf replenishment and ignore restocking costs, e.g., for underfaced items. Chapter 5 therefore extends the basic model by integrating facing-dependent replenishment and inventory holding costs. This chapter is partially based on Hübner and Kuhn (2011a) and Hübner and Kuhn (2011e).

In addition to the classical supply function, shelf inventory has a demand-generating function, as more facings lead to increasing consumer demand. An efficient decision support model therefore needs to reflect space-elastic demand and the logistical components of shelf replenishment. However, shelf space management models have up to now abstracted from reality by assuming that replenishment systems are efficient, and that replenishment costs are not decision-relevant. But shelf space and inventory management are interdependent, e.g., low space allocation requires frequent restocking. Specifically, the models do not differentiate the restocking strategies that retailers frequently use in practice: Scheduled basic group filling by merchandizers of products jointly before sales begin, and concurrent individual product replenishment by sales staff during the sales period.

An extension of shelf space management is therefore provided that additionally takes into account *total costs of overstocked inventory* (TCOI) and *total costs of undersupply* (TCUS).

1.4 Framework for Retail Demand and Supply Chain Planning

$$\text{Max! } P(\bar{k}, \bar{z}) = \text{TDP} + \text{TSP} - \text{TCL} - \text{TCUS} - \text{TCOI} \quad (1.6)$$

TCOI comprise capital costs of overstocked volume $q_i^{(o)}$, i.e., where supply exceeds demand before the next scheduled basic shelf fill process. h is an interest rate and c_i are the product costs. TCUS integrate the additional refilling requirements if demand is higher than supply, expressed by the extra refilling volume $q_i^{(u)}$ and the refilling costs RFC_i.

$$\text{TCOI}(\bar{k}) = \sum_{i=1}^{I} q_i^{(o)}(\bar{k}) \cdot h \cdot c_i \quad (1.7)$$

$$\text{TCUS}(\bar{k}) = \sum_{i=1}^{I} q_i^{(u)}(\bar{k}) \cdot \text{RFC}_i \quad (1.8)$$

Basic constraints are the number of facings, basic supply level and limited shelf space. A multi-product shelf space and inventory management problem is studied that integrates facing-dependent inventory holding and replenishment costs. The non-linear problem is transferred into a knapsack problem that allows fast and efficient solutions using CPLEX. The numerical examples show the benefits of an integrated decision model over traditional approaches and commercial planogram software. Integrating restocking costs aligns facing-dependent demand and cost implications. This results in optimal assortment and shelf configurations from a profit perspective. Sensitivity analyzes are used to additionally compute error bounds for the parameter estimates. Finally, managerial decisions and constraints on operational fulfillment are analyzed as part of a comprehensive hierarchical retail planning framework.

1.4.5 Shelf Space, Assortment and Price Planning

The impact of price on demand and overall profit has so far been excluded in literature on this topic. Chapter 6 now relaxes this assumption and integrates price variations into the decision calculus. Early results are also in the working paper Hübner and Kuhn (2011b).

Traditional shelf management models allocate shelf space to selected items. However, they abstract from the category manager's same-time decision problem to decide not only about space allocation, but also about which products to list and how to price them. Therefore an innovative model is proposed that jointly optimizes assortment, space allocation and prices, viewing category profit as a composite function of price- and space-dependent demand, consumer-driven substitution and price-dependent profit. The innovation resides in integrating pricing decisions and substitution effects with shelf space management.

The interdependent demand for an item i depends on the basic demand, space effects and price effects. Price effects are reflected in the price elasticity ϵ_{il} and the prices r_{il} at price level l and r_{in} the base price of item i.

$$d_i(k_i, r_i) = \alpha_i \cdot (b_i \cdot k_i)^{\beta_i} \cdot \left(1 + \epsilon_{il} \cdot \frac{r_{il} - r_{in}}{r_{in}}\right) \quad (1.9)$$

Total profit additionally comprises *total cross-product profit* (TCPP). The inventory costs from Chap. 5 are excluded to allow unambiguous analyzes of demand effects in comparison to the base model. TCPP are cross-product effects resulting from the cross-product change-over of shoppers due to price and space adjustments. Changes to the facings and price levels impact consumer demand. Additional demand is expressed by DG_i as demand gains, whereas demand losses are expressed by DL_i.

$$\text{Max! } P(\bar{k}, \bar{r}, \bar{z}) = \text{TDP} + \text{TSP} + \text{TCPP} - \text{TCL} \quad (1.10)$$

$$\text{TCPP}(\bar{k}, \bar{r}) = \sum_i^I DG_i(\bar{k}, \bar{r}) - DL_i(\bar{k}, \bar{r}) \quad (1.11)$$

Major constraints are the number of facings, price corridors, basic supply level, minimum volume and limited total shelf space. The model can be implemented as a knapsack problem in CPLEX to provide fast and practical solutions. Numerical examples are used to illustrate insights into planning problems. Sensitivity analyses are used to evaluate error boundaries for consumer behavior and managerial decisions. The analyses show low influence of deviations on estimated consumer behavior, and moderate effects of overarching strategic decisions.

1.4.6 Conclusions and Outlook

The final Chap. 7 summarizes the results and draws conclusion for further research opportunities and implementation in retail practice.

Four avenues emerge as important directions of future research. First, master category planning problems need to be aligned with hierarchical and vertical planning interdependencies. Second, incorporating other consumer demand effects in shelf space optimization models seems a valuable area of research. Third, different modeling techniques and solution procedures may be interesting to cope with large-scale problems that integrate other demand, stochastic or dynamic effects. Finally, most of the existing theoretical models have not been implemented in industry applications (meaning also their theoretical predictions have not been empirically tested). The field would benefit from such applications and empirical tests to validate the assumptions in the increasingly complex shelf space management planning models being formulated in academic literature.

1.4 Framework for Retail Demand and Supply Chain Planning

This research extended the existing literature that addresses the shelf space allocation problem. It does so by capturing the critical decision trade-offs faced by retailers in optimizing their shelf space. This dissertation structures the planning problems, devises decision support models to maximize category profit, and provides methods to test the capability of models for category-specific problems. The planning issues are illustrated using models with differing levels of integration. Specifically, this dissertation develops models that optimize retailers' decisions relating to assortment size, number of facings, replenishment frequency and product prices in a retail category. It would be rewarding to see these models integrated into day-to-day retail practice, with all the benefits that this would imply for both the trade and ultimately for the customer.

Chapter 2
Framework for Retail Demand and Supply Chain Planning

2.1 Introduction

Retail shelf management means matching cost-efficient consumer demand with retail operations. This takes especially place at the point-of-sales in front of the retail shelf, where consumer demand meets retail offer. Therefore, we develop a comprehensive planning framework for retail demand and supply chain planning in this chapter, before we analyze and develop decision support systems for retail shelf management. The increasingly competitive retail environment requires greater customer orientation and operational efficiencies. Consumers are always demanding higher service levels and better purchase prices. Retailers are striving towards broader product variety, better prices and lower costs, and are growing vertically by taking over more logistical functions. Effective structures and planning tools for demand and supply chain planning (DSCP) are therefore the core technique for coordinating thousands of individual decisions in supply chain and customer management.

If consumer demand and retail supply plans are not aligned, retailers need to either solve logistical issues with expensive ad hoc solutions or mark down oversupplied goods (Fisher and Raman 2010). Both approaches lead to a deterioration of the profit base. Consequently, retailers need efficient modeling and decision-making techniques. An analytics orientation at many retail organizations provides a great opportunity for modelers and nudging retail managers towards more quantitative decision-making (Kopalle 2010; Hübner et al. 2010). A comprehensive operations planning framework that integrates consumer behavior is required to maintain and increase retailers' profit both directly (e.g., reduced stockouts) and indirectly (e.g., higher customer satisfaction). While consumer integration into supply chain management (SCM) gains further importance for retailing business practice, only few analytical explanations with consumer integration for retail DSCP have been put forth. Retail practice and research lack a holistic framework that is based on an integral planning perspective at the entire retail supply chain (SC) and comprehensive quantitative decision support systems (DSS).

The goal of this chapter is therefore to derive, structure and integrate holistically DSCP problems and provide quantitative decision support models. This will enable practitioners and researchers to classify decisions and realize the interdependencies of decisions that have to be taken. The framework will help to manage properly the complexity of various planning aspects of interrelated supply chain and category management. As a result, it will foster the transition from research to retail as a comprehensively integrated planning framework. This is the first contribution that structure DSCP planning questions in one framework that matches demand and supply from a long- to short-term perspective and from supplier to customer.

However, due to the lack of literature and since common scientific approaches like structured interviews and questionnaires did not seem to be very promising because a lot of confidence is needed to get this sensitive information, this work is based on experience in retail research and practice. This chapter combines the knowledge and insights into these fields from retail grocery, retail consultancy, science and projects with grocery retailing. Hence, the characterization of the retail planning framework builds on various joint projects with retailers and communications with retail planners and with responsible from consumer goods industry and consultancies. Note that we apply a similar approach as for example in Meyr (2004), Agrawal and Smith (2009c), and Fisher and Raman (2010). Additionally, literature is reviewed to support the conclusions with publications on dedicated planning problems. The research models can both explain and capture the real-life operational processes and decision-making problems, thereby aiming to support decision-making on design, planning, controlling, and executing operations (Bertrand and Fransoo 2002).

The remainder is organized as follows: First, the research objective is specified and corresponding frameworks are reviewed. Section 2.3 sets the context of the decision problems in the retail industry, before Sect. 2.4 formulates a retail DSCP framework and proposes an innovative way for hierarchical retail planning aspects. The Sect. 2.5 illustrates interdependencies between planning modules. The final Sect. 2.6 draws conclusions and develops areas for future research.

2.2 Contribution to Planning Frameworks

2.2.1 Research Objective

DSCP is very complex. While retail managers strive to follow the industry mantra "retail is detail", most retail managers have little time to consider the details of different planning aspects. Also, not every detail that has to be considered at the actual execution can be reflected in the planning process. One core proposal in DSCP is to abstract from reality and to use models as a basis for plans. Analytical models emerge as the most promising solutions to many of the DSCP problems, especially as the advances in computing capabilities allow solving larger problems

2.2 Contribution to Planning Frameworks

(Kopalle 2010). Retail research literature is rich on DSSs for single planning problems (see for example the literature reviews of Levy et al. 2004; Levy and Grewala 2007; Kopalle 2010 or Akkerman et al. 2010). Also Agrawal and Smith (2009b) and Fisher and Raman (2010) describe various retail planning problems. However, they stick to isolated planning problems and do not provide an integrated planning framework or analyze the interdependence of them. That is why it is not surprising, that practitioners often complain about the limited practical value or limited scope of DSS, and challenge the possibility of integrating them into current systems (Kuhn and Sternbeck 2011). Current publications on retail demand and supply chain management do not provide a comprehensive planning architecture, as they deal mainly with isolated planning problems. Fisher and Raman (2010, p.127) note that "Retailers have three tactics at their disposal for matching supply with demand: accurate forecasting, supply flexibility, and inventory stock pilling". We want to broaden this perspective as retail research and practice lack a holistic framework taken into account:

- *Integral planning perspective* at the entire retail SC: A planning framework structures intertwined planning tasks along the SC and takes into account horizontal and vertical interdependencies.
- *Comprehensive quantitative decision support systems*: A planning framework considers proper definition of objectives, constraints, and alternatives. In addition, exact and/or heuristic optimization and evaluation methods are required in the DSSs (Fleischmann et al. 2008).

Our contribution therefore resides in developing a comprehensive operations planning framework that integrates all relevant planning aspects, structures them according to the flow of goods and hierarchical aspects, and is based on quantitative DSS. Researchers and especially practitioners will gain insights as we answer our research questions:

- How should a comprehensive demand and supply chain planning framework be structured?
- How should the planning subsystems be arranged?
- Which planning decisions (in which sequence and hierarchy) are required for a better demand and supply matching?
- And last, but not least: What is the state-of-the art of DSS for each planning module?

Scientist will find further areas of research. We see these especially in a focal literature review for some planning modules that we develop, the development of efficient decision support systems and a unified modeling structure. Retailers and software vendors can take the framework for tailoring, harmonizing or developing advanced planning software systems in retail.

Before developing our DSCP framework, we want to review state-of-the art retail planning frameworks:

2.2.2 Integral Planning Perspective at Entire Retail Supply Chain

The "Efficient Consumer Response" (ECR) concept is an example for integral planning. The initiative aims for a better coordination between strategic partners with the objective to improve satisfaction of consumer needs through efficient replenishment, store assortments, promotions and product introductions (ECR Europe 2003a). This is based on the awareness, that demand-side concepts retain a direct relation to logistical tasks (Kumar 2008). ECR objectives are to remove inefficiencies in information flow and data management along the SC and to realize gains for all partners, which would otherwise not be realized in an isolated, uncoordinated approach (Kotzab 1999). ECR aims towards a vertical cooperation by building trustful relations and sharing data, rather than towards the application of comprehensive analytical methods. The initiatives are "manual"-like qualitative approaches to support the planning. Alvarado and Kotzab (2001) analyze the lack of theoretical explanations of SCM-methods within the ECR. They hypothesize that the complexity involved in SCM leads to difficulties in testing "hybrid" relationships and concepts. A broader theoretical basis appears to be required, especially as "optimization" is an often-used word there. But typically "optimization" conveys there the idea of improvement and not to find the best solution out of a huge, sometimes not countable number of possible alternatives.

On the other side, integral planning fosters to generate total systems, which then cannot be solved optimally because of the complex interdependencies. Varying decision owners, time horizons, frequencies and degrees of aggregation or importance, force a decomposition of decisions (Fleischmann and Meyr 2003). Optimal planning of an entire retail SC is neither possible in form of a monolithic system that plans all tasks simultaneously nor by performing the various planning steps simply successively. The fully integrated optimization could not be put into practice. Also, the poor successive planning would miss optimality. Hierarchical planning is a compromise between practicability, integrating interdependencies and break down the overall planning into partial planning modules (Miller 2001; Schneeweiss 2003a; Fleischmann and Meyr 2003; Gebhardt and Kuhn 2008; Stadtler 2008; Günther and Meyr 2009). It enables coordinating a solution and considering interdependencies. Thus, in the following a hierarchical planning concept is reviewed, which is mainly based on quantitative decision methods.

2.2.3 Comprehensive Quantitative Decision Support Systems

The SC planning matrix from Fleischmann et al. (2008) classifies planning tasks according to the SC processes and the planning horizons (see Fig. 2.1).

They distinguish horizontally along the flow of goods between procurement, production, distribution and sales. Procurement is concerned with providing the resources. Production covers aspects from location planning to shop floor control. Distribution bridges the gap between production and customers. Sales consider

2.3 Specifics of Grocery Retailing

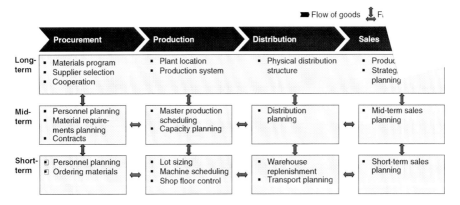

Fig. 2.1 Business-to-Business supply chain planning framework (Fleischmann et al. 2008)

mainly demand forecasts. The vertical distinction of planning tasks into long-, mid- and short-term is determined by the importance and planning horizon.

Fleischmann et al. (2008) group long-term planning tasks into one module to illustrate the comprehensive character of strategic planning. It designs the overall configuration and requires long-term investments in the infrastructure. Mid-term master planning coordinates daily business and typically develops plans for 6–24 months. Finally, short-term planning specifies activities before the execution. The disaggregation of data and options follows the decreasing planning horizon down the hierarchy. The planning modules are coordinated with adjacent modules and account for interdependencies between product and information flow.

The strength of the SC planning of Fleischmann et al. (2008) lies in the decomposition of decisions and structuring planning problems. It aligns planning aspects according to the time-horizons and the value chain activities. However, it has been mainly designed for a business-to-business environment and does not explicitly integrate consumer behavior and store management, which is of major relevance in retail. The sales domain is reduced to product program planning and demand forecast and therefore mainly serves as data input for other planning areas.

The literature so far does not provide a hierarchical and sequential planning concept for the retail industry. Therefore, we structure advanced DSCP tasks, provide a framework for a consumer-driven SC and provide a state-of-the-art overview of DSSs. First of all, we will detail the requirements for grocery retailing. This builds the foundation and defines the requirement for a retail specific operations framework, which follows afterwards.

2.3 Specifics of Grocery Retailing

2.3.1 Attributes of the Retail Grocery Supply Chain

The typology of Meyr and Stadtler (2008) is used to describe the grocery SC functional attributes (*procurement type*, *production type*, *distribution type* and *sales type*) as well as structural types (*topography* and *integration and coordination*). Since typologies can never be comprehensive, this focuses on planning relevant attributes.

Grocery retailers buy and sell standardized fast-moving consumer goods as make-to-stock products. They source different products from many suppliers and have short and reliable replenishment cycles. Demand has to be estimated and may be unstable due to several reasons, e.g., seasonal influences or consumer preferences. Warehouse operations are the production areas where goods are picked for the deliveries to the stores. Distribution can occur in multiple ways from direct store deliveries to multiple distribution centers. Some of the products have dedicated distribution requirements, e.g., a continuous cooled transportation. The retail sales type can be differentiated into products from the permanent assortment and promotional products. Products from the permanent assortment have a stable life cycle compared to other industries and stable prices over a mid-term period. Promotional articles are only temporarily listed and have dynamically varying prices. Both non-seasonal and seasonal products are in retail assortments. With regard to relevant attributes for planning, products are heterogeneous and product sales are high. Major constraints are in capacity like shelf space, outlet backroom, transportation and picking. The SC is inter- and intra-organizational coordinated, as some suppliers maintain own shelf space in retail outlets. Inter-organizational coordination gained increasingly attention by concepts of ECR or Vendor Managed Inventories, particularly these ones that are concerning consumer demand and improving information flows between the SC-members. Quante et al. (2009) classify grocery retail supply chains as flexible, inventory-based chains with make-to-order products.

2.3.2 Reasons for Modifications in Retail Planning

The structural properties of the retail SC are challenges to design a comprehensive DSCP framework. The reasons for modifications in grocery retailing are especially *consumer interaction, focus on sales and demand planning, distribution management* and *warehousing as production*.

2.3.2.1 Consumer Interaction

The key to retail success is to understand and manage the consumer behavior. Fernie and Sparks (2009) highlight the impact of consumers on SC, as retailers can be identified as "active designers and controllers of product supply in reaction to known customer demand." Grewal and Levy (2007) describe it as customer experience management, which includes all points of contact at which the customer interacts with the business, product or service. Blackwell and Blackwell (1999) state that the greatest benefits in SCM can be derived from demand side management. The consumer's behavior initiates many of the activities and processes in the SC (Pal and Byrom 2003). The goal thereby is to satisfy consumer demand quickly, with reasonable quality and at efficient costs and prices. Instore behavior of many consumers determines requirements for retailers DSCP (e.g., in Drèze et al. 1994; Büschken 2009; Chandon et al. 2009; Xin et al. 2009; Hübner and Kuhn 2011e). Furthermore, consumers are part of the value chain, as they execute the final process, taking products from the shelves to the point of consumption (Granzin et al. 1997).

2.3.2.2 Focus on Sales and Demand Planning

The location of the decoupling point has a decisive impact on planning as it divides the planning tasks into forecast-driven and order-driven processes. Incoming orders have to be served at the decoupling point immediately for customer demand, i.e., the crucial decoupling point in consumer goods industry is typically located at the store of the retailer. That means it is at the very end of the supply chain. Thus, forecasting and sales planning gain higher importance in comparison to other industries. Retailers anticipate consumer demand downwards the SC until the "moment of truth" when consumer finally take their instore decisions. Retailers need to proactively manage supply and demand requirements until the consumer enters the store, as all following processes are only reactive. For example, retailers can vary product offers or prices based on anticipated consumer behavior. This refers, not only to demand forecasts, but also to proper optimization methods to steer consumers' behavior. Hence, the sales area influences entire DSCP (Quante et al. 2009).

2.3.2.3 Distribution Management

Primary objective of retail is to bridge the gap between the point-of-production and the point-of-sales at the retail outlet. This means to manage the collection and commissioning of goods in multiple warehouses and the distribution to multiple outlets for up to 50,000 items (EHI Retail Institute 2010). This requires practical planning methods for multi-echelon inventory control (e.g., at one warehouse for n retailers) and distribution planning for multiple nodes, delivery modes and thousands of items with varying transport requirements (e.g., chilled, ambient, fresh) (see e.g., Agrawal and Smith 2009a).

2.3.2.4 Warehousing as Production

Consumer goods are highly standardized products. Picking in the warehouse can be treated as the retailer's production equivalent as store orders are produced in the warehouse. Planning methods are similar to the added value processes of producing industries (e.g., lot sizing and scheduling) and depend on the product characteristics. Nevertheless, the warehouse operations have direct impact on the retail store operations (e.g., if incoming goods need to be re-picked or restored in the store's backroom).

These examples of retail specifics will be reflected in the planning activities. In the following we identify and describe the retail-specific DSCP problems. We distinguish these along the time horizon and the flow of goods, by developing a consumer-backed DSCP framework. The interrelation of planning aspects requires a holistic approach. Horizontally and vertically coherent modules need to communicate to ensure proper material flow and plan alignments. The framework integrates retail specifics, supplier and consumer interaction, as well as hierarchical and sequential decision aspects.

2.4 Framework for Retail Demand and Supply Chain Planning

2.4.1 Overview of Retail Demand and Supply Chain Planning

Large grocery retailers often need to deal with planning decisions on thousands individual items and outlets. Therefore, this section develops a hierarchical set of planning purposes to ensure an *integral planning perspective* at the entire value chain and illustrate the benefit of *comprehensive quantitative decision support systems*. The total system is decomposed into planning modules. Figure 2.2 structures planning problems within the retail DSCP matrix.

Along the SC the domains procurement, warehouse, distribution and sales are distinguished. Planning tasks are classified according to the planning horizon into long-, mid- and short-term. Long-term planning covers strategic decisions. They typically concern configuration decisions of the entire chain. Mid-term master planning coordinates and determines ground rules of the regular operations for the next 6–12 months. It considers also seasonal demand patterns. Retailers normally

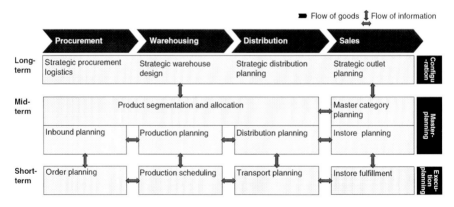

Fig. 2.2 Retail demand and supply chain planning framework – overview

2.4 Framework for Retail Demand and Supply Chain Planning

develop business plans for this period ahead, e.g., when negotiating with suppliers. The short-term execution planning specifies all activities for immediate execution and control within the next days or weeks to react directly on actual product requirements. Individual planning modules are constituted within a matrix to describe these planning problems.

The modules need to be linked by vertical and horizontal information flows. That means, planning results of an aggregated level set restrictions for the subordinate modules, and the results of the lower module communicate back to higher levels. In the following paragraphs, the planning modules from long- to short-term are specified. Furthermore, available scientific models for the planning problems will be appended. Afterwards, the interdependencies between planning modules are illustrated.

2.4.2 Long-Term Configuration Planning

Strategic network and outlet design are coherent strategic problems that cover all configuration areas and define layout structures of the entire network. Network planning includes *strategic procurement logistics*, *warehouse design*, *distribution planning* and *outlet planning*. As a result, network planning constitutes the form, structure and efficiency of the whole retail SC (Gill and Bhatti 2007). It defines the relations between suppliers, retailers and customers (Levy and Grewal 2000). Figure 2.3 summarizes the strategic planning areas for the convergent and divergent product flows.

Long-term configuration planning shapes the entire enterprise, needs to be embedded in the environment and needs to be highly tailored to the retailer's overall philosophy and strategy.

2.4.2.1 Strategic Procurement Logistics

Retail procurement includes purchasing of consumer goods from the supplier and its associated logistics. Consequently it links the retailers' operations with the supplier. It becomes more and more complex and important with an increasing number of

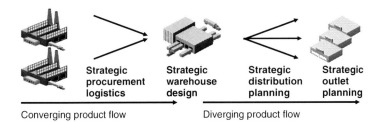

Fig. 2.3 Retail supply chain network

suppliers and their geographical dispersion (Ganeshan et al. 2007). The following three planning areas are distinguished:

The *sourcing strategy* sets the frame for procurement activities by determining the extent of sourcing activities, minimum and maximum number of suppliers per category and sourcing quota, e.g., branded product and private label or single and multiple sourcing. Furthermore, it determines segment-specific sourcing modes and degrees of external service providers in sourcing.

Supplier selection and contracting determine rules for supplier selection and performance assessment, select suppliers based on defined decision criteria and establish contracts. Supplier selection is a multi-criteria decision problem with criteria like price, trade terms, reliability, lead time, producer or brand image. Some factors are hard to quantify, e.g., reliability may play an even bigger role than payment conditions, as out-of-stock situations deteriorate customer satisfaction. Agrawal et al. (2002) determine the optimal suppliers for multiple products with varying demand. They quantify multiple vendor qualifications with premiums like flexibility. Lowson (2001) identifies further criteria like the market power of players. The contracting ensures basic price and quantity agreements, sets the contract duration as well as trade and logistical terms including quantity discounts, minimum order quantities or logistical terms as incentive to improve delivery efficiency. This consists of discounts and agreements, like ex-works or free delivery. Wang and Liu (2007) analyze supplier contracts when the retailer is the dominant partner. Further, Lyu et al. (2010) examine the cooperation between supplier and retailer and propose collaborative replenishment models. This entails a series of planning requirements, such as integration of wholesalers and distributors or joint assortment and supplier selection. These examples already show the extent and complexity of supplier management in retail. The reader is referred to the literature above for further details.

Inbound logistics strategy determines the degree of own inbound SC activities and how these can be organized with suppliers, e.g., if inbound is coordinated by retailer or producer. It covers long-term aspects of the inbound logistics and how goods flow from supplier to retailer and regulates responsibilities. Finally it creates structures for direct store and warehouse deliveries.

2.4.2.2 Strategic Warehouse Design

Warehousing in retail consists of all activities between inbound from suppliers and outbound logistics to outlets, especially picking outlet specific orders as "retail production." The strategic warehouse design rules *number, location, function and types of warehouses*.

The determination of the number of warehouses and location planning requires a trade-off decision between inbound transportation costs, fixed site costs, inventory costs and outbound transportation costs (Teo and Shu 2004; de Koster et al. 2007). It sets the total number and size of warehouses and selects the site. Increasing the number of warehouses reduces the outbound transportation costs, but increases the inbound transportation and inventory costs. The function within the network is deciding whether to use central and/or regional warehouses, e.g., products are stored in a central *and* a regional warehouse or only in a central *or* regional warehouse. Retail particularly operates in various warehouse-outlet modes (e.g., supplier warehouse, retailer warehouse, cross-docks, direct store delivery). The type determination means selecting the technologies for serving the warehouses and running production processes, e.g., technologies for frozen, chilled and ambient products or material handling technologies for order picking systems, like picker-to-parts or parts-to-picker (Dallari et al. 2008). Furthermore, determining the layout structure of the order picking involves a trade-off between travel distance and the required space for the picking area (Roodbergen et al. 2008). The objective of internal design is in the most cases the minimization of handling costs represented by a function of travel distance (de Koster et al. 2007; Gu et al. 2007).

2.4.2.3 Strategic Distribution Planning

The *strategic distribution planning* means determination of the *physical distribution structures* between warehouses and outlets, the flexibility and the degree of own outbound SC activities. The strategic distribution planning creates structures for the transportation mode for direct links to stores, cross-docking or transshipment (Jayaraman et al. 2003). It is a trade-off between infrastructure costs, inventory holding costs, transportation costs and the customer service targets. Often, retailers use a mix of distribution structures depending on the supplier location and the product characteristics. Many retail chains have developed logistics systems for transport between central and regional warehouses and stores. Often, these warehouses are used for storage and picking, but sometimes they only serve for cross-docking. Erlebacher and Meller (2000) minimize the fixed operating and inventory holding costs incurred by warehouses, together with total transportation costs. Hence, it also requires defining global delivery strategies and aspirations for lead times. Van der Vorst (2009) simulates the (re-)design options of a food supply chain. Finally, it needs to regulate to which extent the retailer takes care of the outbound logistics and how strategic partnerships with logistic service providers could be arranged. For a dedicated literature on food distribution planning, we refer to a recent review of Akkerman et al. (2010).

2.4.2.4 Strategic Outlet Planning

A fundamental change in retail pertains to the expansion in the number of retail formats (Kabadayi et al. 2007). The *store type and location planning* includes selecting a set of store types with typical store sizes and determining the outlet network and locations from competitive, customer and logistical perspective (e.g., further geographical expansion vs. increasing network density in established markets). Location is a critical factor in the consumer selection of a store (Cachon and Kök 2007) and requires methods of geographical marketing. Additionally, the outlet has to be easily accessible for supply from warehouses. As a result, outlet planning cannot be planned separately to the distribution planning. Durvasula et al. (1992) presented a model that incorporates managerial evaluations in combination with consumer data. Mendes and Themido (2004) and Grewal and Levy (2007) summarize location decision models, whereas for example Hernandez et al. (1998) and Drezner (2009) develop frameworks for it.

Strategic layout planning determines instore infrastructure and layout at show- and backroom. The showroom layout needs to reflect the retailer's image, must reduce consumer search costs, influences consumers' buying decisions (Drèze et al. 1994; Xin et al. 2009) and impacts space productivity. Finally, the layout impedes instore-logistics processes and sizes the capacity and infrastructure of the backroom storage (Kotzab and Teller 2005). Research models predominately analyze layout planning with empirical studies (Iyer 1989; Turley and Milliman 2000; Lam 2001; Mattila and Wirtz 2008). Hui et al. (2009) developed a probability model for consumer behavior and derived implications for the design of retail space with simulations.

2.4.3 Mid-Term Master Planning

Mid-term master planning utilizes the configuration design of the superior strategic planning and deals with the coordination and planning of aggregated operative units 6–12 months in advance. In the following subsections, firstly the master coordination with *product segmentation and allocation*, and the subordinated mid-term planning problems of *inbound*, *production* and *distribution planning* will be described, before the planning modules for *master category* and *instore planning* of the sales area are constituted that formate the core of retail shelf management.

2.4.3.1 Product Segmentation and Allocation

Planning of product segmentation and allocation is the core of master planning. This planning step coordinates comprehensively the flow of goods between procurement, warehousing and distribution. It comprises of five subproblems.

The *planning of product delivery modes* determines product-supplier-specific delivery modes from inbound logistics to outlets and coordinates inbound, production and distribution planning. These planning tasks comprise decisions about product specific flow types from producer to outlet, e.g., selection between direct store delivery, supplier cross-dock, central and regional warehouses or combinations of it. For example, daily fresh bakery products may be delivered via direct store delivery whereas other non-perishable food products are delivered via warehouses. Thonemann et al. (2005) provide retail-specific criteria for selecting different delivery modes and expose the comparative advantages and disadvantages. This also shows, that it can not be solved independently from the strategic warehouse design.

First of all, *assignment product-warehouse-outlet* builds and assigns products and outlets to warehouses and secondly allocates inventories, e.g., frozen products can be delivered to all outlets from a central warehouse, as requirements for cooled SCs require special equipment. The allocation of inventories across warehouses and outlets builds a liaison across master planning modules. Retailers could hold inventories at central and/or regional warehouses and at the back- and showroom of the outlets. Inventories along the SC are used to increase economies of scale, to balance supply and demand at the individual stations and to hedge against uncertain demand and shortcomings. Since distribution managers are increasingly caught between market pressures on the one hand and the trade-off between high cost of inventories and increased shipment frequencies on the other hand, it is important to plan carefully how centralized a product should be stored (Whybark and Yang 1996).

Selection of dispatch units chooses standardized order units for all products and therefore the granularity of possible order sizes. For example, it is the selection of packaging units (e.g., full carton, full pallet) as shipping units, taking into account handling aspects, inventory aspects, packaging density and shelf capacity. Ketzenberg et al. (2002) explore the instore-benefits of breaking bulk and delivering single units instead of case-packs from the warehouses in the outlets. They quantify the effects on space management and profitability. By employing a simulation model, Yan et al. (2009) consider the impact of case-pack-sizes on the performance of a periodic review inventory system in the presence of spatially-correlated demand.

Building product segments with related order patterns is the aggregation of products to segments with homogeneous order characteristics, including the determination of the associated lead time from outlet perspective. Products could be grouped according to identical order patterns, e.g., homogeneous subsets with respect to shelf life, temperature and other logistical factors (Sternbeck and Kuhn 2010). Furthermore the order lead times for the product groups have to be determined. The shorter the order lead time, the more limited the degree of freedom for order capacity balancing.

Selection of transportation means determines the optimal mix of transportation means (e.g., truck or railway) and desired flexibility within the physical distribution structure. It selects product carriers like returnable boxes, container and pallet for frequent warehouse picking, transportation and shelf replenishment. *Selection of transportation providers* determines mid-term the degree of own transportation services and selection of external providers. That means it covers, for example, the outsourcing of logistical activities to third parties, using own vehicle fleets or third-party carriers (Fernie 1999; Le Blanc et al. 2006; Potter et al. 2007). Hertel et al. (2005) define service, integration, market and cost oriented criteria for logistics outsourcing decisions.

2.4.3.2 Inbound Planning

Inbound planning covers all mid-term problems related to supplier interaction, which concerns supplier order management and master inbound route planning. *Supplier order management* determines order rules, i.e., order quantity planning to define the economic order sizes. It plans dispatching rules for fixed order periods (r, S-policy) or fixed order quantities (s, q-policy) (Tempelmeier 2008). Retailers need to determine optimal order quantities with respect to quantity discounts, time-varying prices (e.g., for promotions), trade terms (e.g., minimum order quantities, delivery costs) and limited storage capacity. Hence the supplier order management also determines the inventory levels and safety stocks. It forces a trade-off decision between the risk of stock outs, obsolete products, high inventory holding costs and dealing with limited storage and shelf space (Ganeshan 1999; Helnerus 2009). Zhao et al. (2004) study a modified economic order quantity problem for a single supplier-retailer system with production, inventory and transportation costs.

Master inbound route planning falls only into the retailers planning domain, if the retailer coordinates the inbound logistics. This planning step aggregates supply points to macro- and micro delivery regions. It coordinates the convergent product flows from suppliers to the retailer, e.g., organization and coordination of producer deliveries or retailer pick ups with milk runs. The objective is the allocation of supplier factories to pick up regions to realize bundling effects. Additionally, master route schedules are determined by region, to realize effects of transportation bundling under constraints like transportation capacity or lead times.

2.4.3.3 Production Planning

Mid-term capacity, personnel planning and warehouse management problems are consolidated in production planning. *Capacity and personnel planning* firstly aligns mid-term storage capacity with a master production program and its forecasted demand requirements. Secondly, it aligns the master production schedule with seasonal fluctuations of demand and calculates a frame for necessary amounts of overtime and additional picking capacity. Examples are the acquisition or rental of additional storage capacity during peak seasons, two- or three-shift models or seasonal engagement plans. Personnel planning covers the optimization of mid-term and general personnel deployment schedules, with respect to employee working time, e.g., sizing of the monthly headcount according to expected capacity requirements. A review of literature for this topic is provided by de Koster et al. (2007).

Warehouse management plans the warehouse layout, allocates storage areas and aligns product flows within the warehouse (Hwang et al. 2004; de Koster et al. 2007). That leads to arrangement decisions on the flow of material and placement within the warehouse in order to optimize storage handling processes. For example, a retail warehouse may mirror a retail outlet, i.e., products are arranged for picking in the warehouse as the store layout (roll cage sequencing). Objectives of storage assignment are high space utilization and efficient material handling (Gu et al. 2007).

2.4.3.4 Distribution Planning

Distribution planning comprises outlet order management and master outbound route planning. It prepares the decision to fulfill the customer service targets with the minimum cost as a trade-off between inventory management policies for each shop and delivery policies from the central warehouse.

Outlet order management determines order rules for outlets. It is, similar to supplier order management, the determination of order policies, with fixed order periods or fixed order quantities. For example, order rules determine delivery frequencies for outlets and product segments with related order patterns. The objective is to minimize delivery costs by achieving high truck utilization and still allowing flexible outlet ordering. Retailers need to optimize delivery frequency with respect to product shelf life and processes in warehousing, transportation and outlets by setting delivery frequency, for example weekly, twice a week or daily for product groups. Ganeshan (1999) studies an inventory policy for a network with multiple suppliers by replenishing a central depot, which then redistributes to a large number of retailers. Viswanathan and Mathur (1997) empirically verify the performance of power-of-two policies under which each retail outlet is replenished at constant intervals. These are power-of-two multiples of a common base planning period. Helnerus (2009) analyses inventory management models for outlets with perishable and non-perishable goods with stochastic demand, fixed capacity and varying assortments.

Master outbound route planning describes planning problems related to transportation links for the outbound-logistics. It is the aggregation of outlets to macro- and micro delivery regions and the determination of master route schedules by region to realize effects of transportation bundling under the constraints of transportation and lead time requirements. The main objective is to minimize costs with regard to high truck utilization, limit empty return transports and further constraints to warehouse and store operations (e.g., time windows at inbound gates) to fulfilling demand on an aggregated level. For example, Pamuk et al. (2004) model the assignment of customers to weekdays and delivery frequency. Adenso-Díaz et al. (1998) also determine on which week day a customer should be served.

2.4.3.5 Master Category Planning

Master category planning frames the sales planning tasks of category management, including overall category sales planning, assortment architecture and promotion planning. *Category sales planning* includes category selection, determining the role of the category and category volume forecasting. Category selection covers determining the set of categories, categories' share and a master product program.

For example, a relative share between adjacent categories could be defined. Overall, it assesses the overall performance of individual categories and determines the mix for an entire store. Gruen and Shah (2000) identify factors that impact categories performance. Category management receives input from the overarching decision of the network and store configuration and sets guidelines for the subordinated sales planning problems.

The determination of a category role defines the role from the consumer perspective, the depth of the category and defines price levels of categories relative to each other. With the beverage category as an example, the retailer needs to decide whether only soft drinks should be listed, or if alcoholic drinks should also be included. Furthermore, the relative price differences between subcategories, such as wine and beer, need to be defined. These planning problems have a predominately qualitative decision characteristic to reflect strategic goals like the price image of a category.

Forecasting category demand generates the mid-term sales plans. It is a crucial component of planning in the retail sector. Inaccurate demand planning affects entire SC performance and behavior. This results in inaccurate material planning, unreliable production schedules, high inventories and stockouts, poor customer service and finally adjustments to sales and marketing projections. This is again a defective input for further demand planning activities with negative consequences for the entire SC. Hence, forecasts drive planning.

Forecasting can normally be derived from historical sales data by adjusting a category's past demand to make projections for the next 6–12 months based on aspects like consumer trends and seasonal influences. Information systems deserve special attention in accurate planning, as retail operations are becoming more complex and forecasts are required on all SC areas. Forecasting is not a retail-specific issue. Only a few contributions deal with retail specific problems. Kumar and Patel (2008) improved the accuracy with clustering products, Bunn and Vassilopoulos (1999) and Dekker et al. (2004) investigated seasonality, whereas Aviv (2001), Yue and Liu (2006) and Aviv (2007) studied forecasts for two-stage SCs. Joint forecasting utilizes and shares information among chain members, i.e., retailer and producer develop forecast plans and resolve replenishment issues together (Levy and Grewal 2000).

The *planning of the assortment architecture* deals with the problem of assortment, shelf space and price management. It is a tactical decision as it implies changes, e.g., for supplier selection and shelf layout (Kök et al. 2009; Fisher and Raman 2010). It receives input from the overarching master category planning. Allocating shelf space and prices is a core problem with special regard to the increasing product variants and the demand for better service levels and prices, and becomes increasingly difficult for retailers.

Assortment management is (de-)listing products for each outlet. When optimizing assortments, it is indispensable to include consumer demand (Anupindi et al. 2009; Smith 2009b). The total demand for a product not only consists of own initial demand, but also the substitution demand from other products (Borin et al. 1994). Mantrala et al. (2009) developed a framework that highlights trade-off decisions, which retailers must make for the assortment planning of how many categories to offer, assortment depth and establishing service levels. Chapter 3 will provide a comprehensive literature review on assortment and shelf space planning in retail. We reflect substitution models for determining the breadth and depth of assortments and optimal inventories as in newsboy models (van Ryzin and Mahajan 1999; Smith and Agrawal 2000; Mahajan and van Ryzin 2001; Kök and Fisher 2007; Yücel et al. 2009).

Shelf space management assigns facing quantities to individual products with limited shelf sizes and restocking capacity. Planning models are provided for the quantity of inventory that should be carried out for each item (Urban 1998; Abbott and Palekar 2008; Hübner and Kuhn 2011e,a), the amount of space that is assigned to each product (Hansen and Heinsbroek 1979; Corstjens and Doyle 1981; Hübner and Kuhn 2011d) and its location within the store (Yang 2001; Hariga et al. 2007). Hansen et al. (2010) compare heuristics for decision models with facing-dependent demand and vertical and horizontal location effects.

The price management sets prices relative to other product prices and to competition. Prices of the permanent assortment are usually set for mid-term periods. They need to be adjusted continuously due to mid-term variances in demand, competition, seasonality or costs of operations (Grewal and Levy 2007). However, pricing and assortment decisions are strongly intertwined (McIntyre and Miller 1999). Murray et al. (2010) model jointly shelf space and pricing decisions. Retailers often use mixed calculation with loss leaders in the assortment. This results in low margins for the discounted items, but profit and demand gains for other items if consumers shop for these together. An integrative approach treats demand planning and pricing as part of the consumer goods SC, as decoupling takes place at the point of sales. Consequently, retail DSCP methods need to be extended to demand fulfillment and revenue management (Quante et al. 2009). For example, Hall et al. (2010) analyze retailer pricing and ordering decisions in a dynamic category management setting.

Promotion planning determines promotional type, assortment and prices, and evaluates effectiveness. Promotion management focuses on promotional items and plans, whereas assortment architecture optimizes the permanent sections in the store. Promotions studies evaluate the success and optimize temporary promotions (Gupta 1988; Abraham and Lodish 1993; Blattberg and Neslin 1993; Federgruen and Heching 1999; van Heerde et al. 2004; Smith 2009b; Fisher and Raman 2010).

2.4.3.6 Instore Planning

Instore planning arranges store staff and store logistics concepts. The *store personnel planning* aligns the store personnel resources (in regular terms, e.g., quarterly), with requirements such as expected customer frequencies, customer service and replenishment activities. The staff acquisition in retail is similar to other industries, however, the restrictions differ, such as assignments for entire opening hours or restocking during night shifts. Retailers face a trade-off between hiring too many employees, facing high costs, but also assuring a minimum service level in the store (Lam et al. 1998; Thonemann et al. 2005; Kabak et al. 2008).

Store logistics planning determines logistical instore and shelf replenishment processes. Relevant planning questions are for example dealing with refilling quantities, cycles and time windows for shelf replenishment. Objectives are to reduce stockouts and meet required outlet service levels, increase logistical efficiencies and free up time for customer service. This can be achieved e.g., by outsourced merchandisers or a scheduled refilling before store opens. Further examples are analyses of instore refilling processes and the determination of a level for re- and intermediate storage of goods in the backroom (DeHoratius and Ton 2009; van Donselaar et al. 2010; Fisher and Raman 2010). It receives input from the shelf space planning and the operational constraints form the subordinated instore fulfillment. Kotzab and Teller (2005), Broekmeulen et al. (2006), Helnerus (2009), van Zelst et al. (2009) and DeHoratius and Ton (2009) describe instore logistic and replenishment models.

2.4.4 Short-Term Execution Planning

The short-term execution planning develops action plans on an hourly to weekly basis for actual requirements and jobs.

2.4.4.1 Order Planning

Order planning includes order dispatching, route planning and ramp management for actual material and order needs from suppliers. *Supplier order dispatching* determines operative order quantities and times for actual material requirements with respect of supplier delivery frequencies, trade and logistical terms as well as storage capacity. It answers the short-term question when and how much to order. Order dispatching solves the trade-off between inventory and order costs, i.e., order costs increase with frequent orders, whereas inventory costs increase with infrequent orders.

When the retailer coordinates the inbound logistics, *operative inbound route planning and transport scheduling* allocates actual retail orders to detailed transportation capacity (e.g., available trucks). Furthermore it develops a time-phased deployment schedule and determines operational resources like staff and vehicles for actual pick up jobs (Fleischmann and Meyr 2003). For example, it bundles pick ups across suppliers to achieve higher truck utilization while meeting lead time limits and determines daily route schedules.

Inbound ramp management allocates delivery slots and coordinates incoming goods. Illustrative examples are the length of warehouse opening hours for incoming logistics or queuing and sequencing at the inbound gate.

2.4.4.2 Production Scheduling

Production scheduling deals with lot-sizing as well as assignment of machine and personnel resources within the warehouse for actual orders. *Personnel scheduling* determines time plans and responsibilities on a weekly or daily basis to ensure fulfillment of picking activities. For example, it requires an alignment of weekly staff employment plans with expected outlet order volume and other warehouse jobs (Kabak et al. 2008).

Lot sizing is the core decision at warehouse operations as it (dis-)aggregates actual outlet order to delivery units. It is the picking of multiple goods according to outlet demand. It is necessary to build lot-sizes for balancing the costs of inventory holding and changeovers with respect to picking and transportation capacity, as well as lead times. Changeover costs increase with small, frequent lot sizes, inventory costs with large, infrequent lot sizes (De Koster et al. 1999; Bozer and Kile 2008). Further planning questions are technical requirements, e.g., use of mixed pallets.

Sequencing and job release deals with the final assignment and scheduling of individual orders to resources and employees. For example, it determines schedules by reflecting lead times and due dates, delivery requirements, utilization rate and available capacity, transportation and inventory holding costs. And finally, the job release determines timing and start dates for actual jobs, e.g., time schedules for individual picking jobs. Gu et al. (2007) discusses comprehensively operative warehouse decisions.

2.4.4.3 Transport Planning

Transport planning plans the fulfillment of product flows between locations for actual outlet orders.

Outlet order dispatching determines operative order quantities and times for actual outlet material requirements with respect to delivery frequency, distribution and storage capacity (e.g., to meet service levels) (Le-Duc and de Koster 2007; Bozer and Kile 2008).

The *operative outbound route and transport scheduling* allocates actual outlet orders to detailed transportation capacity to develop a time-phased deployment schedule. It also assigns operational resources like personnel and vehicles for actual distribution jobs, i.e., it determines the operational resources necessary for transporting products located at a central facility (e.g. a distribution center or a warehouse) to geographically dispersed retail stores (Anily and Bramel 1999). For example, it combines transports for a single outlet across different category orders or assigns orders to deliveries according to the master route planning guidelines. This also includes vehicle loading, i.e., adjusting the number of shipments with multiple items on the same transport link to full truck loads (Fleischmann and Meyr 2003). Cardós and García-Sabater (2006) analyze the dependence of transport planning, inventory management and consumer response on service levels.

The *outbound ramp management* allocates distribution slots and coordinates outgoing material flows. Similarities with inbound ramp management include opening hours, queuing and sequencing.

2.4.4.4 Instore Fulfillment

The final short-term planning deals with the instore fulfillment of operative tasks, i.e., plans short-term sales, develops personnel schedules, and aligns operative instore logistics and shelf replenishment processes. Instore fulfillment at the "moment of truth" significantly influences the performance of a retail store. For example, Angerer (2006) shows that instore processes cause 72% of the out-of-stock situations, and only 28% have their origin outside the outlet. Broekmeulen et al. (2006) and Kuhn and Sternbeck (2011) concluded that the instore logistic costs amount up to 48% of the total logistic costs.

Short-term sales planning forecasts individual product demand and adjusts inventory levels and prices. The short-term sales forecasting details the category forecast on a daily to weekly basis for individual products according to expected short-term effects and other category planning decisions. It enables customer demand fulfillment from shelf stocks (available-to-promise) for each listed product. Moreover, the short-term alignment of inventory levels and prices for perishable products could result in mark-down prices in order to sell them before the date of expiry. The dynamic pricing and inventory control can be planned with revenue management approaches (e.g., Federgruen and Heching 1999; van Donselaar et al. 2006; Smith 2009a). Gallego and van Ryzin (1994) identify it as a problem, where the seller owns perishable goods that are sold to price sensitive consumers with time-varying prices. Hamister and Suresh (2008) conclude that dynamic pricing reduces sales variability, and the corresponding bullwhip effect, and increases the profit compared to static pricing in retail. Their single-item newsboy problem optimizes stock and price levels. Federgruen and Heching (1999) developed an inventory model with temporary price variations. They established optimum period pricing and stock levels. Yin et al. (2009) combine dynamic pricing with shelf space allocation, whereas Minner and Transchel (2010) determine dynamic order quantities for perishable products with limited shelf-life.

Store personnel scheduling determines personnel schedules and responsibilities on a weekly or daily basis to ensure fulfillment of operative instore logistics and customer activities. For example, it aligns personnel availability with customer frequency and rush hours. It needs to define the operations time windows for each individual store employee, e.g., shift time, time allocation to shelf replenishment, customer service or cash desk, and must be aligned with customer frequency (see also Fisher and Raman 2010). Lam et al. (1998) studied a personnel planning model, where they use store traffic forecasts to determine the number of employees. Kabak et al. (2008) determined the optimal headcount with a sales response model, and then they assigned employees to daily schedules and finally established a revision process between the planning stages. Zolfaghari et al. (2007) developed heuristics for shift assignment and scheduling for retailers.

Short-term instore logistics management plans the physical flow of goods in the stores to the shelves for actual delivery and refilling needs. It includes short-term planning of refilling management and sequencing of shelf replenishment jobs. The refilling management determines the frequency for shelf inventory level control and degree of re-picking and re-packaging in the backroom for example. Shelf replenishment is a major cost driver of the entire SC (Thonemann et al. 2005), therefore reducing handling processes comes to the fore. Optimization of instore logistics does not only start in the stores, but already at the picking at the warehouses. Outlet-specific commissioning allows streamlining processes in the outlets, as handling costs can be reduced. The sequencing determines how and in which order incoming goods flow from outlet ramp to shelves, e.g., how many goods get simultaneously replenished, in which order and how priority rules for shelf replenishment can be established. Hariga et al. (2007) and Abbott and Palekar (2008) are modeling optimized shelf replenishment cycles.

However, the available literature on the planning for instore logistics management is limited, as it focuses mainly on causal correlations (van Zelst et al. 2009). Wong and McFarelane (2007) developed systematic means for describing different shelf replenishment policies. McKinnon et al. (2007) analyzed root causes for out-of-stock situations in forecasting, ordering and replenishment processes. Waller et al. (2008) investigated the impact of the selected case-pack size on backroom logistics and store-level fill rates. Van Zelst et al. (2009) identified case pack size, number of case packs replenished simultaneously, the filling system and the employees as the most important efficiency drivers. All authors provide estimates for the respective effects, but do not provide true optimization. This may be due to the fact that short-term requirements for outlet inventory replenishment is mainly customer-driven and thus finally exogenous, hence it requires master planning concepts as provided above to anticipate consumer demand and replenishment requirements.

2.4.5 Summary of the Retail Demand and Supply Chain Planning Framework

The operations framework structures DSCP problems for grocery retailing as in Fig. 2.4 and Fig. 2.5. Also it builds the global framework for shelf management. Single planning modules are developed and supplemented with literature on quantitative models where feasible and applicable. The matrix provides a comprehensive guide for retailers DSCP problems and serves for the classification of planning problems and based on these it helps to discover untapped potential. The framework fosters the transfer of scientific knowledge to retail practice, as it clarifies planning

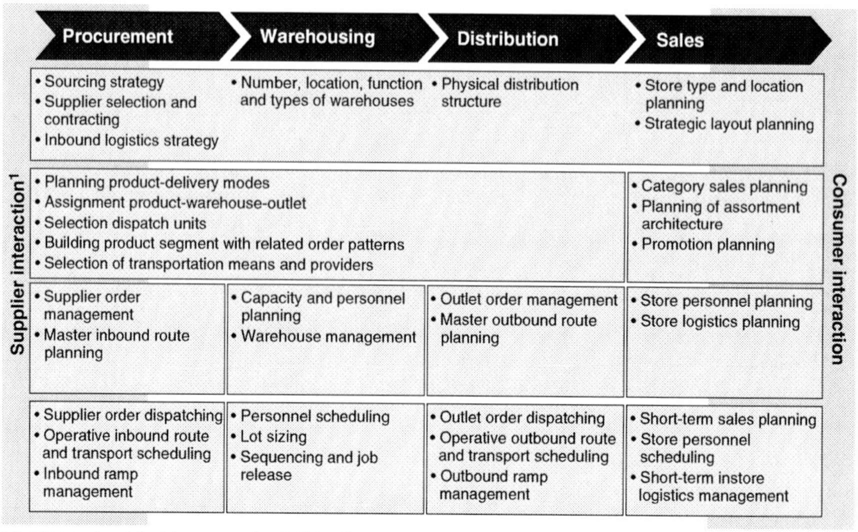

	Procurement	Warehousing	Distribution	Sales	
Supplier interaction[1]	• Sourcing strategy • Supplier selection and contracting • Inbound logistics strategy	• Number, location, function and types of warehouses	• Physical distribution structure	• Store type and location planning • Strategic layout planning	**Consumer interaction**
	• Planning product-delivery modes • Assignment product-warehouse-outlet • Selection dispatch units • Building product segment with related order patterns • Selection of transportation means and providers			• Category sales planning • Planning of assortment architecture • Promotion planning	
	• Supplier order management • Master inbound route planning	• Capacity and personnel planning • Warehouse management	• Outlet order management • Master outbound route planning	• Store personnel planning • Store logistics planning	
	• Supplier order dispatching • Operative inbound route and transport scheduling • Inbound ramp management	• Personnel scheduling • Lot sizing • Sequencing and job release	• Outlet order dispatching • Operative outbound route and transport scheduling • Outbound ramp management	• Short-term sales planning • Store personnel scheduling • Short-term instore logistics management	

[1] including supplier-relationship management and collaboration

Fig. 2.4 Retail demand and supply chain planning framework – detailed view

processes. Moreover, it highlights "white areas" where additional quantitative models may help to improve planning processes.

Retail planning ranges from supplier to shopper. Consequently, the framework is embedded into supplier interaction upstream the SC and needs to integrate consumer interaction at the point of sales downstream the SC. The supplier-relationship management and collaboration deals with long-term to short-term planning of joint initiatives and communication channels, e.g., electronic data interchange, vendor managed inventories or joint product development. For example, recent literature examines the role of producers and retailers in product innovation (Ganeshan et al. 2007). Also, Groothedde et al. (2005) study the prospects to develop a collaborative hub network, aiming to consolidate material flows between manufactures and retail distribution centers. Aydin and Porteu (2009) analyze the effects of manufacturer-to-retailer versus manufacturer-to-consumer rebates in a supply chain. Kurtulus and Toktay (2009) and Kurtulus and Toktay (2011) evaluate the role of category captainship practices.

On the other hand, the presented matrix incorporates consumer behavior. The consumer interaction at the point-of-sales affects the entire planning process. Retailers need to integrate consumer trends, behavior and reactions into overall operations planning. It requires an integrative approach that treats consumer demand planning as part of the entire operations framework. Especially as decoupling is located at the outlets, retail SC methods need to include also demand planning, demand fulfillment and revenue management concepts (Quante et al. 2009). This will be the focus in the following chapters.

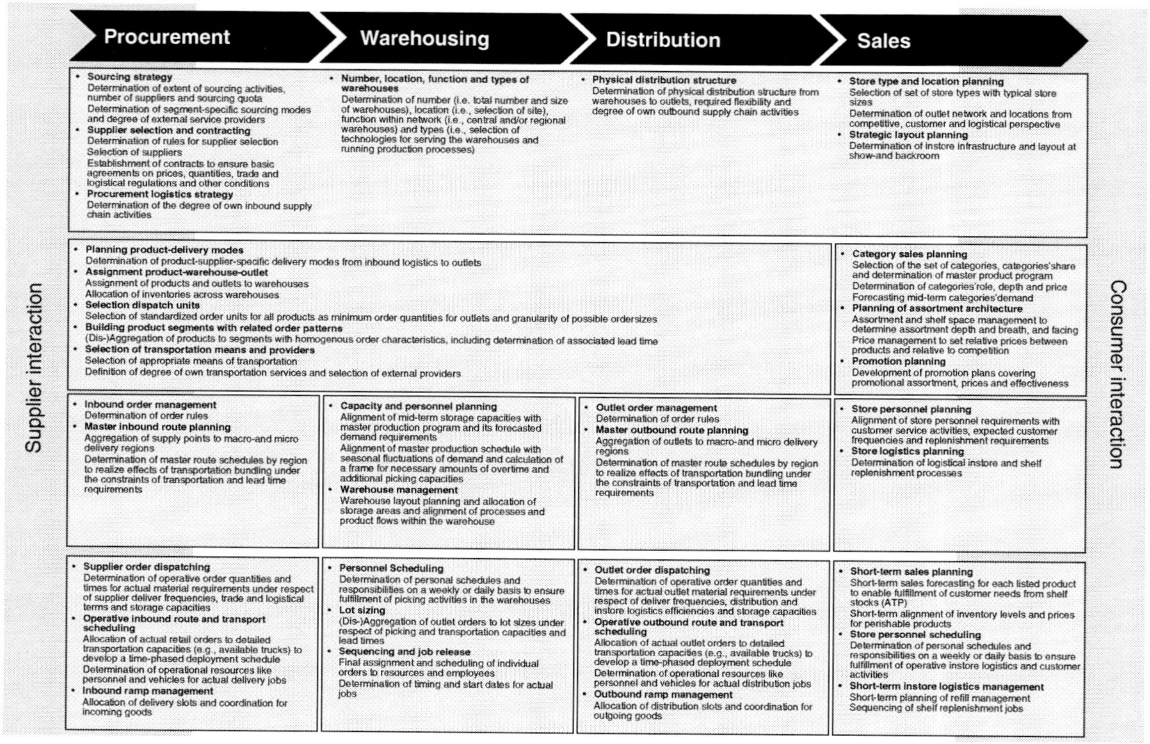

2.4 Framework for Retail Demand and Supply Chain Planning

Fig. 2.5 Retail demand and supply chain planning framework – summary

2.5 Aspects of Planning Interdependencies at Retail Shelf Management

After constituting individual planning modules, now sequential and hierarchical interdependencies along planning modules with quantitative approaches will be illustrated for retail shelf management.

Sequential planning means that planning processes should consider the SC from consumer to supplier as an entire system and taking into account the interdependencies of various planning activities. *Hierarchical planning* describes planning processes that consider the hierarchy of decisions and integrate them where advantageous and feasible. *Comprehensive quantitative decision support models* ensure tightly aligning retail shelf supply with consumer demand and leading to efficiency gains for the entire chain. Some examples will illustrate this. The first ones describe a sequentially integrated retail logistics example, the second type hierarchical interdependencies.

2.5.1 Example: Vertically Integrated Shelf Space and Price Management

Retail outlet sizes and formats are determined on a long-term basis in the *store type and location planning*. This sets the boundaries for a strategic layout planning of how the instore infrastructure and the layout at all can be determined. The master category planning then takes decisions on overall category selection, forecasting individual category sales and plans assortment architecture and promotions. For example, a retailer needs to decide how many categories should be listed in a store, and how much space each should retain. Similarly, but on a more granular level the decisions take place for individual products. The decisions are arranged in an hierarchical sequence, but could potentially be solved integrated as well. A decomposition of the decisions into subproblems and communication flows between sub-planning modules aligns the planning decision and information flows.

Examples of vertically integrated models are the following ones: Yin et al. (2009) combine the planning problems of dynamic pricing and shelf space allocation. They provide a game-theoretical model for a retailer with a limited inventory of a product over a finite selling season. The authors show that the selection of a certain type of inventory display format has an impact on the optimal markdown pricing strategy. Another example is provided by Hariga et al. (2007). They introduce a model for the determination of the optimal shelf space allocation and replenishment. The objective is to decide on the optimal number of facings and replenishment time in order to maximize the retailer's profit.

Similarly, the information flows need to take place between lower modules, e.g., the forecasted sales need to be an instruction for store personnel planning and store logistics planning.

2.5.2 Example: Horizontally Integrated Retail Operations Form Warehouse to Shelf

The planning modules need also be aligned along the supply chain. For example, mid-term master planning needs to reflect interdependencies between warehouse, distribution and sales for the determination of the order patterns. Often retailers operate different flow types from producers to the stores (e.g., direct store delivery or central warehouse delivery). Products are segmented and allocated to flow types by specific logistical factors. Each segment requires the determination of an order pattern (e.g., delivery every Monday – Wednesday – Friday). While retailers set often segment-specific order patterns by simple rules (e.g., the higher the sales volume of the store the higher the delivery frequency), analytical DSSs enable to derive an outlet and segment specific cost minimum by focusing on the relevant processes in the sequential areas of warehousing, transport and outlet (Sternbeck and Kuhn 2010).

A further example for horizontally integrated planning is Bhattacharjee and Ramesh (2000). They examine a multi-period retailing problem for a monopolistic retailer dealing with a product with limited product life cycle. A model is presented that optimizes dynamic pricing and ordering policies jointly. Tellis and Zufryden (1995), as well, provide a model, which integrates dynamic pricing and ordering by determining the optimal timing and depth of retail discounts.

These examples show the need for communication flows and alignment between individual planning modules. The communication flows in form of instructions, anticipations and feedback loops will be part of the further analyses in this work.

2.6 Conclusions and Future Areas for Research

The objective of the retail DSCP matrix is to give a general outline of the main DSCP problems that are relevant for the majority of grocery retailers. This framework can help to improve retailers profitability with the proposed "more analytics based thinking" (Kopalle 2010, p.118). This chapter provides a comprehensive retail operations planning framework consisting of dedicated and coherent planning modules. The framework structures planning problems and provides a state-of-the-art overview of retail DSCP. The examples show the need to balance integrated planning and true optimization with hierarchical and sequential planning aspects. Further research opportunities are in the general application, the development of specific DSSs and the implementation of advanced models in commercial software.

2.6.1 General Application of Retail DSCP Matrix

First of all, the list of DSCP problems in the architecture is potentially not complete. Many retailers might be confronted with lots of other planning details depending on their individual situation. Additionally, planning problems and horizons may differ in specific cases depending on the structural type. For example, if the marketing strategy forces to change the store layout every few weeks, store layout planning turns to a short-term problem. Secondly, the framework has been developed for grocery retailing, but would need to be tailored to other consumer-facing industries, e.g., fashion or pharmacies with specific planning requirements. Variations in the SC networks, degree of SC control, geographical spread, relative logistics costs, level of data and IT, as well as relative sophistication of service providers may result in various focus areas (Fernie and Staines 2001). The concept should be generic enough to cope with these variations.

2.6.2 Unified Modeling Structure

The architecture fosters the understanding of both horizontal and vertical planning problems. However, it has also been shown that the implementation of hierarchical planning structures can be difficult in practice. Therefore, modeling the relationship between hierarchical and adjacent levels is one of the main research opportunities for implementing decision support systems (Zoryk-Schalla et al. 2004). An analytical hierarchical planning process will need to be developed and tailored to our framework. A unified modeling structure would help to analyze the coordination of the different levels and the interdependencies in decision-making (Schneeweiss 1998; Miller 2001; Schneeweiss 2003a,b; Stadtler and Kilger 2008; Günther and Meyr 2009). Thus, Chap. 3 will provide a unified modeling structure for retail shelf management.

2.6.3 DSSs for Dedicated Planning Problems

It has not been anticipated that the exposition of research topics is comprehensively exhaustive. However, the mentioned literature indicates that the amount and type of literature is different for each planning problem. For retail-specific planning problems such as assortment planning, shelf space allocation or dynamic pricing, there is a great amount of retail specific literature available. Other planning problems like demand forecasting and product ordering are deeply investigated, but since it is not only industry specific, the models are less focused on retailing and the challenges there (e.g., non-stationary demand, seasonality, unobservable lost sales). Future research could concentrate on developing further DSSs for problems with currently less advanced models and by integrating retail specifics. We will develop a capacitated assortment and shelf management model in Chap. 4.

2.6.4 DSSs for Interrelated Planning Problems

Future research areas concern the hierarchical and sequential planning, and the use of comprehensive quantitative DSSs. The advances in IT and computation technologies encourage enlargement of hierarchical planning over the entire SC processes (Fleischmann and Meyr 2003). This requires a discussion of the possibility of decomposition and coordination of decentralized decision making and analyzing the distribution of decision rights and information asymmetries within the SC. The following areas illustrate ideas for developing more sophisticated models. Considering the hierarchy of decisions would be beneficial for example in evaluating interdependencies between store layout, master category and assortment planning, or between shelf layout and replenishment. Furthermore, considering the SC from consumer to supplier as an entire system and taking into account the interdependencies of various planning activities would be advantageous, e.g., for connecting supplier selection with assortment management or joint optimization of order patterns and shelf layout. We will analyze the effects of integrated shelf space and inventory planning in Chap. 5 and extension with pricing decisions in Chap. 6.

2.6.5 Implementation of Advanced Models in Commercial Software Packages

The planning modules of the framework need to be transferred into software modules for practical decision making. Practitioners favour user-friendly and simplistic software tools. Here science and implementation show a big discrepancy. Science focuses on the approach to integrate extensive interdependencies resulting in complicated estimation requirements of parameters. Advances in scientific modeling still need to demonstrate their implementation ability. The main barriers and areas of investigation to retailers' adoption are requirements for large item sets, data availability, model complexity, difficulty of integration into existing systems or cross-functional interfaces. We will therefore also review software applications for retail shelf management in the following chapter.

Altogether, this shows that DSCP is very complex, but offers a lot of research opportunities. The purpose of this chapter was to structure the problem for researchers and retailers. By this, it will foster the development and dissemination of DSS to retail practice. In the following chapter, we analyze empirical insights, quantitative models and software applications for master category planning. We will focus on the mid-term sales domain of the RDSCP-matrix.

Chapter 3
Empirical Insights, Quantitative Models and Software Applications for Master Category Planning

3.1 Introduction to Master Category Planning

Several mutually reinforcing trends have made category management (CM) one of the most critical marketing and operational decisions for retailers. German retailers and consumer goods producers recently rated "optimization of product portfolio and category management" the most important task for achieving performance goals (Breuer et al. 2009). This is not surprising as shelf space competition in retail stores is at an all-time high, driven by the competitive need to constantly introduce new products. There has been significant product proliferation since the 1990s (Greenhouse July 17, 2005). The average number of items in overall store assortments increased by 30% between 2000 and 2009 (EHI Retail Institute 2010). In confectionery, for example, the number of brands rose by more than 40% between 1997 and 2001, but overall volume by only 0.8% (Carlotti et al. 2006). Additionally, most retailers suffer from decreasing space productivity. Gutgeld et al. (2009) concluded that 19 out of 24 European retailers were unable to maintain their space productivity.

The increasingly competitive environment, the growing need for operational efficiencies and ever greater customer orientation are therefore forcing retailers to develop efficient decision support systems (DSS) for category planning. Retailers are required to make a large number of mid-term CM decisions that include, above all, which categories and products to offer, shelf space allocation, inventory levels and replenishment cycles. Retailers need to resolve the conflict of ever-increasing number of consumer goods with scarce shelf space and the high operational costs relating to great product variety (Grocery Manufacturers Association et al. 2005; Desrochers and Nelson 2006; Gutgeld et al. 2009). Offering broader assortments thus may limit the appropriate service levels and vice versa. However, the continually increasing number of consumer goods is in conflict with the fixed and scarce resource of shelf space. Furthermore, as shoppers increasingly take purchase decisions instore, retail marketers are diverting a growing proportion of their marketing budgets from traditional out-of-store media advertising to instore marketing (Chandon et al. 2009). Customers, additionally, have become more exacting, demanding ever-increasing service levels, greater product choice and lower prices (Agrawal and Smith 2009b). Despite heavy investment in point-of-sales scanners and IT, retailers are still losing potential revenue due to their inability to get the right goods to the right places at the right time (Zenor 1994; Basuroy et al. 2001; Dhar et al. 2001; Friend and Walker 2001; Cannondale Associates 2003; Angerer 2006; Fisher and Raman 2010). The focus of this research is therefore to examine retail category problems and DSSs that particularly enable retailers to identify which products to offer, allocate shelf space, and determine restocking frequencies. These planning questions are inevitably interdependent. For example, large assortments drive lower inventory levels of individual items, which reduce the items' visibility on the shelves, increase the risk of stockouts and impose high restocking costs due to the need for frequent replenishment (Campo and Gijsbrechts 2005; Curseu et al. 2009; Mantrala et al. 2009). Consistent with previous research, these planning questions are defined as:

- *Assortment planning*: Listing decisions based on consumer choice behavior and substitution effects.
- *Shelf space planning*: Facing and replenishment decisions based on space elasticity effects, limited shelf space and operational restocking constraints.

It will be shown that these planning problems are not yet sufficiently and comprehensively integrated into commercial software systems and scientific models. Substitution effects for unavailable items are investigated in *assortment planning* models. *Shelf space planning* models deal with inventory- and facing-dependent demand, but largely do not integrate substitution effects for items that are not in the assortment. Most shelf models do not factor in latent consumer demand for non-listed items. The commonly proposed method of handling delisted items is to simply assume lost sales and no consumer substitution (see e.g., Hansen and Heinsbroek 1979; Corstjens and Doyle 1981; Abbott and Palekar 2008 and Hansen et al. 2010). On the other hand, assortment models omit space elasticity effects and to a large extent shelf space constraints, too (Kök et al. 2009).

The goal of this chapter is to identify and structure these relevant shelf planning problems and their associated demand effects in mid-term master CM. It provides a summary of empirical insights, commercial software systems and advances in scientific modeling. These contributions are subsequently examined in relation to each other, outlining discrepancies and areas of research.

The remainder is organized as follows. The second Sect. 3.2 sets the context of CM. Section 3.3 identifies the commercial software applications, while Sect. 3.4 specifies research progress in assortment and shelf space planning. The concluding Sect. 3.5, provides an outlook on further research areas and the discrepancies between scientific modeling and retail practice.

3.2 Definition and Scope of Master Category Planning

We provide a comprehensive framework for the discussion, as most retail category planning takes place from a functional perspective, i.e., item selection and space allocation are made with limited regard to overarching strategies, cross-functional implications or subordinated domains (Griswold 2007). Category planning is a tactical decision as it leads to mid-term changes, e.g., in supplier selection and shelf layout. It receives input from strategic planning and provides instructions to operational instore fulfillment tasks (see Chap. 2). Master category planning entails a series of hierarchical aspects:

Category sales planning includes a selection of categories, and demand forecasting for each. Category selection covers determining the set of categories, each category's share, and a master product program (see also Fisher and Raman 2010). CM receives input from the overarching decisions of store configuration, and sets guidelines for subordinated planning problems.

Assortment planning involves (de-) listing products. When optimizing assortments, it is essential to reflect consumer demand. The total demand for a product consists not only of its own initial demand, but also the substitution by and complementary demand for other products.

Shelf space planning assigns facing quantities to individual products under the constraints of limited shelf sizes and restocking capacity.

Instore logistics planning is investigated as a dependent problem of assortment and shelf space planning. Replenishment frequency impacts the entire shelf supply and actual stock levels, which then influence consumer choice at the shelves.

Figure 3.1 guides the discussion, maps the supply- and demand-based interdependencies, contributes to an integrated management approach, and creates interesting opportunities for transferring insights from one domain to another.

Most retail category planning takes place from a functional perspective, i.e., item selection and space allocation are made with limited regard to overarching strategies, cross-functional implications or subordinated domains (Griswold 2007). This shows that efficient DSS in CM are required to improve business performance.

3.3 Software Applications for Master Category Planning

3.3.1 Popularity of Software Systems in Category Planning

Category planning can be a complex decision process, potentially involving thousands of items requiring an integrated retailer, manufacturer and consumer perspective. Adequate IT systems should allow the gathering, preparing and analyzing of category-relevant information and enable the category manager to evaluate the category's performance. With improving data mining and computation technologies, software vendors are able to offer more category planning solutions, some of which

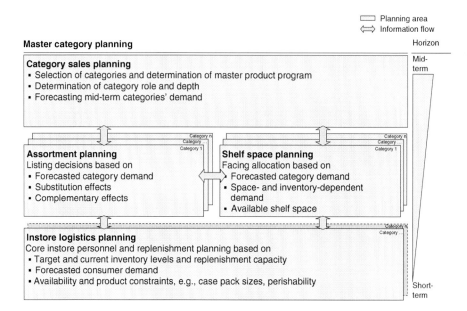

Fig. 3.1 Interdependencies in master category planning

replace homegrown spreadsheet calculations with dedicated software applications (Syring 2003; Griswold 2007; Mantrala et al. 2009; Retail Information Systems et al. 2009). Hence, advanced analytical tools should allow higher decision quality to be achieved.

Large enterprise software vendors provide comprehensive solutions focusing on workflow and data integration, while specialized software companies concentrate on a particular task in the planning process (Becker et al. 2000; Syring 2003; Griswold 2007). The focus in this subsection is on functional capabilities rather than workflow and data management, as the most important applications can be integrated in IT systems like SAP Retail or Oracle (Becker et al. 2000; Griswold 2007). Furthermore, these systems do not comprise self-contained shelf and assortment tools (Becker et al. 2000).

CM software applications offer a wide spectrum of productivity gains in handling large data sets and deciding on multi-item problems with easy-to-understand rules. However, the simplicity of decision making logic may not be sufficient to motivate retailers to invest in such a technology. This requires a closer look at the use of such IT systems in practice. Historically, only a limited share of retailers have used IT support in CM. The picture of IT support for retail CM shows large regional differences. A.C. Nielsen (2004) reports that nearly all US retailers uses CM tools. 86% of US retailers use assortment planning tools and 81% apply shelf management tools, whereas in Europe only a limited proportion of retailers use IT support in CM. ECR Europe (2003a) and Schramm-Klein and Morschett (2004) found in 2004 that 64% of European retailers did not use any IT for assortment planning, and 69%

3.3 Software Applications for Master Category Planning

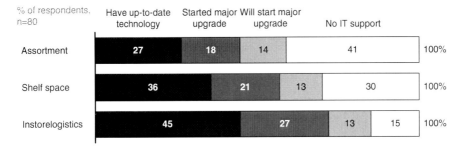

Fig. 3.2 Retailers management capabilities in merchandising (2009)

did not use any IT for shelf space management. More recently, Retail Information Systems et al. (2009) reported a higher degree of IT use and planned investments:

For assortment planning, 45% have already up-to-date software in place or currently upgrading, 14% have plans to invest. 57% have and currently implement shelf space allocation tools, and also 13% will invest. 72% have and implement replenishment planning tools and 13% of the retailers will invest here (Fig. 3.2).

This shows firstly that even through IT support for CM is more common at retailers nowadays compared to 2004, a significant share of retailers – 15–41% – still do not use IT support at all. Secondly, it shows that around 1/8 of retailers plan to invest in CM technology. It also reveals that the more operational the task, the more likely they are to use IT support for their CM. Retailers still partially rely on their own homegrown spreadsheet calculations for the more strategic task of assortment planning.

3.3.2 Scope and Overview

The focus of IT support in CM lies in decision support for individual actions and tactics. Retailers and producers use commercial DSSs to create assortment and merchandising plans. The popularity of these applications is mainly due to their simplicity for implementing decisions on a large number of items. The applications of leading vendors are similar in their purpose, scope and decision logic, as they have their origin and historical focus on shelf space planning. The applications mainly differ in terms of support functions for data exchange, graphical interfaces and output, and reporting functions. Table 3.1 provides an overview of available applications based on Griswold (2007). Figure 3.3 outlines the match between focused planning problems and featured vendor functionalities (Griswold 2007). The strongest features of the vendors' applications are highlighted.

In the following, the focus of our analysis will be on the leading three vendors in assortment and shelf space planning. The functionalities of the tools of AC Nielsen, MEMRB/ IRI, and JDA will be described.

Table 3.1 Overview of major CM software based on Griswold (2007) and own analyses

Area	Vendor	Application	User[a]
Assortment	AC Nielsen	Product planner	>2,000
	MEMRB/IRI	Apollo assortment planner	>800
	JDA	Efficient item assortment	>100
Shelf space	AC Nielsen	Spaceman suite	>2,000
	JDA	Space planning	>2,000
	MEMRB/IRI	Apollo professional	>800
	SAS Institute	SAS Space planning	>170
	AVT/Oracle	Retail focus merchandiser	>65
	Galleria	Space analytics	>30
Other[b]	AC Nielsen	Store planner	>2,000
	JDA	Floor planning	>200
	SAS Institute	SAS StoreCAD plus	>170
	Visual Retailing	Visual retailing	>90
	AVT/Oracle	Retail focus	>45
	Galleria	Store optimizer	>30
	Torex retail	Compass planning and VM	>25

[a] Number of companies using application
[b] Mainly store and floor planning applications

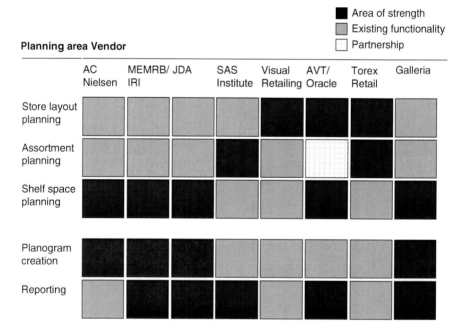

Fig. 3.3 Retail category planning applications – vendors' functionalities and strengths

Assortment planning tools like "Product Planer," "Apollo Assortment Planner" and "Efficient Item Assortment" support item selection using cumulated customer penetration, which attempts to increase assortment coverage. The tools analyze the extent to which modification of assortment plans impact consumer spending or revenue increases. The tools lead the user through the decision process and structure qualitative decisions on assortment planning.

Shelf planning applications like "Spaceman Suite," "Apollo Professional" and "Space Planning" visualize shelf planograms and support determination of the facing quantity and shelf position. By applying large-scale data-processing techniques to existing sales and inventory, they serve to forecast future demand needs at a store and item level. They turn defined assortments into merchandising standards. The main purpose is to simulate alternative item placements on-screen and the related profit analyses. The shelf arrangement can be reproduced and displayed at the screen. The decision logic is based on key performance indicators (e.g., sales, profit) or a mixture of several of them. The tools allocate shelf space according to simple heuristics like proportional-to-market share or proportional-to-profit share. Restrictions can be imposed such as minimum and maximum facings, shelf availabilities in days, or service levels. The software visualizes shelf arrangements of automatically generated planograms. The items are displayed graphically on the screen, where either schematic or photorealistic views can be chosen. Placing an item on the merchandise plan is effected by manual user selection. The software systems provide various reporting functions, evaluate the outcomes, and assist in defining the actions that will produce the best expected results. Consequently, most retailers "use them mainly for planogram accounting purposes so as to reduce the amount of time spent on manually manipulating the shelves" (Drèze et al. 1994, p.14).

All the applications are quite similar and allow the user to obtain successive solutions, from simple shelf arrangements to automated shelf planograms. They create realistic real-time pictures of the assortments without time-consuming computation processes (Becker et al. 2000; Syring 2003; Lehnert and Hüffner 2006; Griswold 2007).

The implementation of shelf management technology has led to significant benefits with revenue increases of 10–20% through reallocation of space to higher velocity items, refined assortments that streamline store-stocking practices and improve in-stock positions (Griswold 2007; Campillo-Lundbeck 2009).

3.3.3 *Limitation of Commercial Software Applications*

Commercial models are limited in their use of mathematical optimization and incorporation of relevant consumer demand effects. No "real" optimization takes place in the sense of profit-maximizing shelf allocation (Zufryden 1986; Syring 2003; Irion et al. 2004; Hansen et al. 2010). The automatically generated shelf recommendation based on a proportional rule is subsequently manually adjusted. A heuristic approach involving a targeted, iterative search for the optimum can only be conducted manually by repeatedly creating planograms. The concern for simplicity and practicability in commercial applications results in suboptimal solutions (Yang and Chen 1999). As mentioned in Levy et al. (2004), a basic weakness of these simplistic rules is that they have nothing to do with the optimal decision. Rather, these rules base their conclusions on what has been done in the past (Kopalle 2010). A second drawback results from the failure to incorporate consumer-based demand effects of shelf space on individual product sales and substitution between products (Drèze et al. 1994; Irion et al. 2004). Griswold (2007, p.1) concludes that: "While software vendors intend to build integrated assortment and space planning functionality, progress is still needed before a complete functional footprint can be developed to support the integrated business process required by retailers." Most CM vendors strive toward comprehensive solutions with integrated planning of store type and layout, assortment, shelf space and planograms. "However, retailers have not bought into the viability of a single application. Because of this, the trend is best-of-breed point application selection, with retailers accepting the burden of integration" (Griswold 2007, p.9).

Progress from research and retail is still needed to produce sufficiently integrated solutions. Most shelf management vendors acknowledge the importance of various planning concepts such as store type and layout planning, assortment planning, space planning and planogram creation, plus maintenance. Retailers' desire for more capable tools could be satisfied by leveraging advanced mathematical modeling techniques and the power of IT to efficiently and profitably manage categories. We therefore review in the following section advances in scientific assortment and shelf management follows.

3.4 Scientific Models for Master Category Planning

Existing research addresses assortment and shelf space planning in different literature streams. Assortment planning is gaining increasing attention in the operations research community due to recent publications on substitution models, whereas shelf space planning has already been on the research agenda for the last three decades (Table 3.2).

We will undertake our literature review from these two perspectives. Retail shelf space models deal mainly with space-dependent demand, but largely do not integrate substitution effects for items that do not fit onto the shelves. Consumer choice effects for non-available items are investigated in substitution models. Substitution models have a predominately stochastic nature, whereas shelf space models are deterministic.

Table 3.2 Core criteria of literature streams in assortment and shelf space planning

Criteria	Assortment planning	Shelf space planning
Substitution effects	X	(X)
Space-elastic demand		X
Limited shelf space	(X)	X

X: fully integrated; (X): partially integrated

3.4.1 Assortment Planning Models

Assortment planning considers the question of which and how many different products to offer (Mantrala et al. 2009). Kök and Fisher (2007) define retail assortment planning as the process used to find the optimal set of products to be carried and set the inventory levels of each product. One main feature of assortment planning is the integration of consumers' willingness to accept a substitute when the favorite product is not available.

3.4.1.1 Empirical Studies of Assortment Planning

The review of assortment optimization models is preceded by a summary of empirical insights. Consumer reaction to assortment size remains unclear. On the one hand, choosing among larger assortments might be demotivating for consumers. There is empirical evidence that variety levels have become so excessive that reducing variety significantly increases sales (Sloot and Verhoef 2008). Iyengar and Lepper (2000) suggest that consumers are more likely to purchase consumer goods when offered a limited array of choices only. "When people have "too many" options to consider, they simply strive to end the choice-making or deal by finding a choice that is merely satisfactory, rather than optimal" (Iyengar and Lepper 2000, p. 999). Drèze et al. (1994) indicate that product variance has reached a saturation level. Dhar et al. (2001) conclude in a study with 19 categories in 106 US supermarkets that reductions in the assortment in staple categories result in higher unit sales relative to competitors. Boatwright and Nunes (2001) found that significant item reductions (up to 54%) resulted in an average sales increase of 11% across 42 categories examined, and sales growth in more than two-third of these categories. On the other hand, Broniarczyk et al. (1998) found that consumers' perception of assortment attractiveness increases with the expectation that the preferred items are stocked, the more items are carried, and the more overall space is allocated to the category.

Even if retailers could determine the optimal assortment mix for all individual customers, it may be unprofitable to stock such an assortment. Therefore out-of-shelf situations are a key reality, which needs predictions for consumer reactions. Campo et al. (2003) point out that out-of-stock (OOS) substitution can be used as an approximation for out-of-assortment (OOA) substitution. They report that although the retailer losses in case of a OOA may be larger than those in case of an OOS (e.g., no postponement, permanent store switching), there are also significant similarities in consumer reactions and OOS reactions can be indicative of OOA responses. Table 3.3 summarizes studies on substitution behavior. For example, Gruen et al. (2002) and Corsten and Gruen (2003) examine consumer response to stock-outs in a meta-study across eight categories at retailers worldwide. They report that 45% of consumers substitute, i.e., buy one of the available items from that category, 15% delay purchase, 31% switch to another store, and 9% do not buy anything.

Table 3.3 Overview of empirical studies on substitution behavior

Authors by year	Consumer behavior, percent of consumer responses				
	No purchase	Post-ponement	Store switching	Substitution	
				Within brand	Other brand
Walter and Grabner (1975)	–	3%	14%	19%	64%
Schary and Christopher (1979)	19%	11%	48%	5%	17%
Emmelhainz et al. (1991)	–	13%	14%	42%	32%
Verbeke et al. (1998)	–	20%	21%	57%	
Campo et al. (2000)	4%	49%	3%	44%	
Campo et al. (2000)	2%	30%	2%	66%	
Zinn and Liu (2001)	–	15%	23%	62%	
Gruen et al. (2002)	9%	15%	31%	19%	26%
Sloot et al. (2005)	3%	23%	19%	18%	36%
ECR Europe (2003b)	9%	17%	21%	16%	37%
Helnerus and Müller-Hagedorn (2006)	–	25%	9%	36%	30%
van Woensel et al. (2007)	–	6%	10%	84%	
Helnerus (2009)	–	48%	30%	7%	15%

[a] Cereals
[b] Margarine

The average potential for substitution depends on product, situation and consumer characteristics (Helm and Stölzle 2009; Helnerus 2009). Fitzsimons (2000) concludes that substitution behavior is driven by the personal commitment to the non-stocked item and the difficulties of switching initial choice. Xin et al. (2009) deduce that sold-out products influence the consumer decision process twofold. First, the situation creates a sense of urgency that impulses consumers to immediately purchase an alternative ("immediacy effect"). Furthermore, sold-out products can enhance the perceived attractiveness of substitutes to the sold-out product. However, these studies confirm also that substitution behavior are immanent and need to be reflected in assortment optimization.

3.4.1.2 Optimization Models for Assortment Planning

Even if retailers could determine the optimal assortment mix for all individual customers, it may be unprofitable to stock such an assortment. Therefore out-of-shelf situations are immanent. Studies show that between 22% and 84% of demand can be substituted. These consumer reactions need to be reflected in assortment optimization.

Rigorous analytical models have emerged as the most promising solutions to many of these planning problems (Agrawal and Smith 2009b; Fisher and

3.4 Scientific Models for Master Category Planning

Raman 2010; Kopalle 2010). The main body of literature on assortment models based on substitution effects focuses on developing algorithms to define inventory levels by stochastic demand. The most advocated solution procedure is based on the Newsboy model. The most popular approaches for estimating demand substitution in assortment planning are *multinomial logit models* (MNL) and *exogenous substitution models*. Only these two model types have been used in recent publications on assortment planning. The MNL is a discrete consumer choice model that assumes that consumers are rational utility maximizers and derives anticipations of consumer behavior. It is commonly used in economics and marketing literature. Similarly, locational choice models are also utility-based models. On the other hand, exogenous demand models directly specify the consumer reaction and are mostly used in inventory models. Because of these structural differences and different origins, the exogenous substitution models and MNL models will be analyzed separately in the following subsections. A basic model of each stream will be presented, followed by a review of related studies and further developments.

A. Exogenous Substitution Models

Exogenous substitution models specify the demand directly for each product. Consumers choose from an item set $N = \{1, 2, \ldots, i, \ldots, j, \ldots, I\}$ and if the preferred item j is not available for any reason, an individual consumer might accept another item i as a substitute according to a defined substitution probability μ_{ji}.

The probability of substituting from item j to i is specified by a substitution probability matrix.

$$\mu_{ji} = \begin{pmatrix} 0 & \mu_{12} & \ldots & \mu_{1i} & \ldots \\ \mu_{21} & 0 & \ldots & \mu_{2i} & \ldots \\ \vdots & \vdots & \ddots & \vdots & \ldots \\ \mu_{i1} & \mu_{i2} & \ldots & 0 & \ldots \\ \vdots & \vdots & \vdots & \vdots & \ddots \end{pmatrix} \qquad (3.1)$$

The parameter μ_{ji} makes it possible to differentiate between items with varying substitution rates. The exogenous demand models generally allow only one round of substitution, and if the first alternative is also not available, the sales are consequently lost (Smith and Agrawal 2000; Netessine and Rudi 2003; Kök and Fisher 2007; Kök et al. 2009; Yücel et al. 2009). Kök (2003) shows that this is not too restrictive.

Smith and Agrawal (2000) developed a probabilistic demand model capturing substitution and defining inventory levels with given shelf constraints. The multi-item Newsboy inventory model sets the stocking level of each product to achieve exogenously determined service levels for sequentially arriving consumers to maximize total expected profit. The probability function $g_i(\bar{z}, C)$ describes that item i is demanded by the Cth arriving customer, either as an initial preference or as a substitute, depending on the set of listed products indicated by vector $\bar{z} = (z_1, z_2, \ldots, z_i, \ldots, z_I)$. The binary variable z_i denotes whether a product i is listed (set to value 1) or delisted (value 0). The first term in (3.2) denotes the probability f_i that an arriving customer will initially prefer item i. The second term describes the probability that the customer will substitute their demand for item j. The probability that item j is available when the Cth customer arrives is expressed by $a_j(\bar{z}, C)$. This probability can be derived from a discrete, transient Markov process formulation.

$$G_i(\bar{z}, C) = f_i + \sum_{j \in N, j \neq i} f_j \left[1 - z_j a_j(\bar{z}, m)\right] \mu_{ji} \quad \forall i \in N \quad (3.2)$$

The model updates the inventory levels after each customer arrival. The terms depend on the choice of previous customers and the number of substitution attempts made by the customer. Since the determination of G_i and a_j is complex, Smith and Agrawal determine a lower bound $L_i(\bar{z})$ for the demand probability. The lower bound is achieved by considering only assortment-based substitution. That means the lower bound is equal to the probability that an arriving customer initially prefers item i and the substitution gains from delisted products ($L_i(\bar{z}) = f_i + \sum_{j \in N, j \neq i} f_j \left[1 - z_j\right] \mu_{ji}$). They use the lower bound for the demand calculation on the assumption that the number of arriving customers per cycle follows a negative binomial distribution. The probability distribution $\psi_i(d)$ determines the item is demand including substitution effects, where $d = 0, 1, 2 \ldots$ denotes a random variable of the demand.

The optimization problem is to maximize the expected category profit

$$\widetilde{\Pi}(\bar{q}, \bar{z}) = \sum_{i \in N} z_i \left[\widetilde{\pi}_i(q_i, \bar{z}) - \text{FCR}_i\right] \quad (3.3)$$

by determining the item stocking quantities q_i, given by the vector $\bar{q} = (q_1, q_2, \ldots, q_i, \ldots, q_I)$, and the set of listed products, given by vector $\bar{z}$. Fixed replenishment costs FCR_i occur when i is included in the assortment. The expected product profit $\widetilde{\pi}_i$ is expressed as a Newsboy profit:

$$\widetilde{\pi}_i(q_i, \bar{z}) = (r_i - c_i) \lambda_i(\bar{z}) - c_i \sum_{d=0}^{q_i} (q_i - d) \psi_i(d | l_i(\bar{z}))$$

$$- (r_i - c_i) \sum_{d=q_i}^{\infty} (d - q_i) \psi_i(d | l_i(\bar{z})) \quad \forall i \in N \quad (3.4)$$

The profit consists of gross profit from realized sales minus salvage costs and lost sales. The item's revenue is r_i, the item's unit costs are c_i, and q_i is the optimal Newsboy stocking quantity to achieve the target stocking level. $\lambda_i(\bar{z})$ denotes the expected demand for item i given a certain set of listed items, denoted by $\bar{z}$.

3.4 Scientific Models for Master Category Planning

Smith and Agrawal (2000) propose an approximate solution with Lagrangian relaxation using the space constraint. This is followed by a complete enumeration for small item sets to define optimal stocking levels, and a linearization approximation for large item sets. Illustrative examples are used to obtain several insights, such as that substitution effects can reduce the optimal assortment sizes, ignoring substitution effects decreases profit significantly, and it may not be optimal to stock the most popular item as a result of varying substitution relationships.

Rajaram and Tang (2001) analyze the impact of product substitution on retail merchandising with a service-rate heuristic based on the basic Newsboy model. They evaluate the impact of substitution on order quantities and expected profits, and show that substitution reduces shortages and overstocking costs. Kök and Fisher (2007) also use an exogenous stochastic model and develop an estimation approach using sales data from different stores. They propose an iterative optimization heuristic for the joint inventory and assortment problem with out-of-assortment (OOA) and out-of-stock (OOS) substitution assuming constraints of shelf space, maximum inventory levels and order lead times. They integrate lost sales arising from unmet demand into their objective function. Shelves are divided into facings, but do not account for space elasticity. Shah and Avittathur (2007) study multi-item inventories based on a Newsboy problem and propose a heuristic for assortment decision and stocking levels. Yücel et al. (2009) model an assortment and inventory problem under consumer-driven demand substitution. They conclude that neglecting consumer substitution, excluding supplier selection or ignoring space limitations all have significant impact on the efficiency of retail assortments.

B. Multinomial Logit Substitution Models (MNL)

MNL models use consumer preferences. These models have their origin in economics and marketing literature, and have recently been applied to assortment and inventory planning as well.

Van Ryzin and Mahajan [1999] study a static assortment planning approach, where the demand derives from stochastic choice processes in which individual purchase decisions are made according to a MNL random utility model. Their objective is utility maximization in the setting of a Newsboy problem. They assume identical costs c and identical prices r for all items. Listed items are denoted by set N^+ and delisted items by set N^-. Thus, $N^+, N^- \in N$, $N^+ \cup N^- = N$ and $N^+ \cap N^- = \emptyset$. In addition, there is a no purchase option, denoted by $i = 0$. Thus, $N_0 = N \cup \{0\}$ contains all possible options. A customer chooses the variant with the highest actual utility U_i among the set of available choices, $\{U_i, i \in N_0^+\}$. However, the utility values are uncertain and vary from customer to customer due to heterogeneity of preference among customers. The values are therefore random with an expected utility value of u_i, $i \in N_0$. Consumer preferences v_i can then be defined by $v_i = e^{u_i/\delta}$, $\forall i \in N_0^+$. The positive parameter δ reflects the heterogeneity among the customers, with a higher value of δ corresponding to a higher degree of heterogeneity.

The probability that a customer purchases item i can then be calculated as follows: $A_i(N^+) = v_i/(\sum_{j \in N^+} v_j + v_0)$, $\forall i \in N^+$. If an additional item $l, l \in N^-$, is added to N^+, with $N_l^+ = N^+ \cup \{l\}$, the denominator increases by v_l. Hence, the probability that a customer will select one of the original items in N^+ decreases. However, the overall probability ($\sum_{j \in N_l^+} A_i(N_l^+)$) that a customer selects any item from set N_l^+ increases. The authors assume that customers make their product choice when they observe the assortment, and they are not faced with OOS. Hence, $A_i(N^+)$ is independent of the inventory level. The expected profit $\tilde{\pi}_i$, $i \in N^+$ is

$$\tilde{\pi}_i(N^+) = (r-c) A_i(N^+) \cdot \overline{C} - \text{CF}(A_i(N^+) \cdot \overline{C}) \quad \forall i \in N^+ \quad (3.5)$$

where $\overline{C}$ is the mean number of customers visiting the store per period. Furthermore, the cost function CF(.) is assumed to be concave and increasing to reflect economies of scale. The total expected category profit $\tilde{\Pi}(N^+)$ is the sum of item profits. The optimal assortment is a balance between including a new product that impacts total category demand, but also increases the assortment cost. If the profit gain of product l is more than the sum of the profit losses of the products in N^+, then adding l improves the profit. The profit impact ($\Delta \tilde{\Pi}$) of adding an item l to N^+ is

$$\Delta \tilde{\Pi}(N_l^+) = \tilde{\pi}_l(N_l^+) - \left[\sum_{j \in N^+} \tilde{\pi}_j(N^+) - \sum_{j \in N^+} \tilde{\pi}_j(N_l^+) \right] \quad \forall l \in N^- \quad (3.6)$$

The first term is the expected profit from $l \in N_l^+$, and the term in brackets accounts for the cannibalization losses and profit gain by complementary effects due to adding product l to the assortment. They use a quasi-convex function. Since a quasi-convex function achieves its maximum at the end points of the interval, the profit gets maximized either by adding the product with the highest utility or by not adding a further product to the assortment. Hence, the optimal assortment is always in the popular assortment set (Smith and Agrawal 2000; Cachon et al. 2005; Kök et al. 2009). This reduces the number of assortment combinations to be considered from 2^N to N.

The model considers OOA substitution and has been extended by OOS substitution in a dynamic model of Mahajan and van Ryzin (2001). They propose a stochastic sample-path optimization method for multiple rounds of substitution. The models capture the trade-off decision between larger assortments and increased average inventory costs. Major drawbacks are that shelf capacity is not considered and they use only identical prices and costs for all items. Hence, the theorem of "optimal assortment = popular assortment" does not hold when products have varying prices, costs, demand, product sizes or order quantities.

Cachon et al. (2005) extend van Ryzin and Mahajan (1999) and Mahajan and van Ryzin (2001) with consumer research costs. Annupindi et al. (1997) describe methods for demand estimates and substitution rates, and model consumer arrivals

3.4 Scientific Models for Master Category Planning

in the context of vending machines. They derive maximum-likelihood estimates of arrival and substitution rates via the expectation-maximization algorithm by treating OOS as missing data. Hopp and Xu (2008) study dynamic demand substitution analyzing inventory, price and assortment decisions in competitive scenarios. Gaur and Honhon (2006) study consumer preferences based on a locational choice model. They assume that under static, assortment-based substitution, consumers make decisions independent of assortment availability. In contrast, in dynamic substitution consumers purchase the highest-ranked product available at the time of their visit.

Conclusions from Review of Assortment Models

Table 3.4 summarizes relevant assortment planning models with substitution effects for the retail grocery industry, which analyze product variety and inventory levels together.

Table 3.4 Overview of literature on assortment models

Authors (by year)	Demand model[a]	Model enhancements	Substitution reason	Solution methods	Items[b]
van Ryzin and Mahajan (1999)	MNL		OOA	Specialized heuristic	8
Smith and Agrawal (2000)	ED	Limited space	OOA	Lagrange relaxation, one-dimensional search	5
Mahajan and van Ryzin (2001)	MNL		OOA, OOS	Stochastic gradient algorithm	10
Rajaram and Tang (2001)	ED	Order quantities	OOS	Service rate heuristics	7
Cachon et al. (2005)	MNL	Consumer search costs	OOA	Specialized heuristic	8
Gaur and Honhon (2006)	LC	Location	OOA	Specialized heuristic	5/2
Kök and Fisher (2007)	ED	Limited space	OOA, OOS	Iterative heuristic	29
Shah and Avittathur (2007)	ED		OOS	Specialized heuristic	3
Li (2007)	MNL	Store traffic	OOA, OOS	Specialized heuristic	5
Hopp and Xu (2008)	MNL	Price, competition	OOA, OOS	Fluid networks heuristic	3
Yücel et al. (2009)	ED	Supplier selection, limited space	OOA	MIP models (CPLEX)	10

[a] ED exogenous demand, MNL multinomial logit, LC locational choice
[b] Maximum number of items considered in test case

The exogenous demand model allows more degrees of freedom than the MNL model. Since the options in the set of choices are assumed to be homogeneous, the MNL model cannot capture the various types of adjacent, one-product, or intra-subgroup substitution. The substitution rates in the MNL model depend on the relative utility of the options. This is both an advantage and a disadvantage for the MNL model. The advantage is that it allows one to easily incorporate other marketing variables such as prices and promotions into the choice model. The disadvantage is that it cannot differentiate between the initial choice and substitution behavior. Unlike the MNL model, the exogenous demand model differentiates between categories that have the same initial demand for the category but different substitution rates. Thus, for example, the MNL model cannot treat assortment-based and stock-out-based substitutions differently. In contrast, it is possible to use a different η_j or different substitution probability matrices for assortment-based and stock-out-based substitutions in the exogenous demand model (Kök et al. 2009).

The major criticism of the MNL model results from its Independence of Irrelevant Alternatives (IIA) property. "This property holds if the ratio of choice probabilities of two alternatives is independent of the other alternatives in the choice process (...) IIA property would not hold in cases where there are subgroups of products in the choice set such that the products within the subgroup are more similar with each other than across subgroups."(Kök et al. 2009, p.109) If for example the brand loyalty is high in an assortment with two products from different brands, adding a new sub-brand to the first brand will probably cannibalize the sales of the first brand more than the other second brand (Kök et al. 2009). The introduction of Coke Zero as a brand supplement to the Coca Cola products is such an example.

IIA does not allow to integrate this important aspect of consumer choice. The example most used in the literature to explain this property is the "blue bus/red bus paradox." The example describes a person who has the same probability f of using his car or of taking the bus: $f\{car\} = f\{bus\} = 1/2$: We assume that now there are two buses available that are identical except for their color, red or blue. Further, we suppose that the person is indifferent about the color of bus it takes: The choice set is {car, red bus, blue bus}. We would intuitively expect that $f\{car\} = 1/2$ and $f\{red\ bus\} = f\{blue\ bus\} = 1/4$: However, the MNL model implies that $f\{car\} = f\{red\ bus\} = f\{blue\ bus\} = 1/3$ (Kök et al. 2009).

The advantage of the exogenous demand models is that the substitution probability matrix can take different forms to represent different probabilistic mechanisms. The following examples illustrate the different forms of a four-product category (Smith and Agrawal 2000; Kök et al. 2009).

A *random substitution* as in (3.7) allows random values for the substitution probabilities μ_{ji} for item substitution from j to i.

3.4 Scientific Models for Master Category Planning

$$\mu_{ji} = \begin{pmatrix} 0 & 0.1 & 0.2 & 0.7 \\ 0.4 & 0 & 0.4 & 0.2 \\ 0 & 0.7 & 0 & 0.3 \\ 0.8 & 0.1 & 0.1 & 0 \end{pmatrix} \quad (3.7)$$

The *subgroups substitution* matrix in (3.8) allows for substitution within the subgroups only. For example, in the coffee category, consumers may treat decaffeinated coffee and regular coffee as subgroups and not substitute between subgroups.

$$\mu_{ji} = \begin{pmatrix} 0 & 0.5 & 0 & 0 \\ 0.5 & 0 & 0 & 0 \\ 0 & 0 & 0 & 0.6 \\ 0 & 0 & 0.4 & 0 \end{pmatrix} \quad (3.8)$$

The *adjacent product substitution* matrix as in (3.9) assumes that products are ordered along an attribute space and allows for substitution between neighboring products only. For example, if a customer cannot find 1% milk in stock, they may be willing to accept either 2% or skim, but not whole milk.

$$\mu_{ji} = \begin{pmatrix} 0 & 1.0 & 0 & 0 \\ 0.5 & 0 & 0.5 & 0 \\ 0 & 0.5 & 0 & 0.5 \\ 0 & 0 & 1.0 & 0 \end{pmatrix} \quad (3.9)$$

Retail assortment studies provide various heuristics for consumer-driven substitution. The exogenous demand and MNL models have different origins, but allow the optimization of assortments. MNL models are more stylized from the perspective of integrating consumer choice behavior, and allow structural properties of the optimal solution to be obtained. The core assumption is that items can be arranged by their popularity and the most popular items are in the assortment. Smith and Agrawal (2000) show, however, that it is not always essential to list the most popular item. The exogenous demand models are more flexible, enabling the depiction of more realistic inventory problems involving varying prices or pack sizes for example. The studies have contributed to various directions, such as supplier selection and pricing (see also Khouja 1999; Kök et al. 2009; Pentico 2008; Mantrala et al. 2009).

The average number of items used in the test cases of the proposed model is very limited (≤ 29 items). However, grocery categories have an average of 60–80 items within a category, meaning the practical relevance is limited (EHI Retail Institute 2010). Furthermore, only Kök and Fisher (2007) and Yücel et al. (2009) include scarce shelf space in the considerations, which is a dominant factor for assortment considerations in the retail grocery. Yücel et al. (2009) prove that space limitations have a significant impact on the performance of assortments. None of the studies analyzed take into account the space elasticity effects of a higher number of facings.

Thus the following review will further focus on shelf space models with available shelf space as a restriction, and with space-dependent sales effects.

3.4.2 Shelf Space Planning Models

An underlying assumption of shelf space management is that grocery shopping behavior is susceptible to retailers manipulation. Better shelf display influences shopper behavior, since the majority of consumers decides about their final purchases in the store (Hoch and Deighton 1989; Drèze et al. 1994; Xin et al. 2009; Chandon et al. 2009). Consumers also exhibit a low level of involvement with their instore decisions and make choices very quickly after minimal search (Hoyer 1984).

3.4.2.1 Empirical Studies of Shelf Space Allocation

A series of experimental studies quantified the relationship between demand and facings, with a higher number of facings resulting in higher visibility, consumer awareness and consumer demand. Shopper surveys (Inman et al. 2009) and field experiments (Hansen and Heinsbroek 1979; Corstjens and Doyle 1981; Zufryden 1986; Bultez et al. 1989; Drèze et al. 1994; Koschat 2008; Chandon et al. 2009) conclude that a significant relationship exists between space allocation and demand. The degree of significance depends on the type of items. Brown and Tucker (1961) recognized increasing space effects from (1) the group of unresponsive, price-inelastic products, to (2) general products for everyday purchases to (3) impulse purchases. Curhan (1972) proved that fast-moving goods have a higher space effect than slow-moving items.

Experiments for estimates of space elasticity deliver a "rules of thumb" of 0.2 as space factor, i.e., doubling the facing led to a 20% increase in sales (see for example Drèze et al. 1994; Campo and Gijsbrechts 2005). Drèze et al. (1994) were able to prove a 4% profit increase through better managed shelf space, although with diminishing returns for further increasing space. Desmet and Renaudin (1998) calculate category space elasticity of up to 0.8. Their results show increasing space elasticity with impulse buying rates, independent of the type of store. Koschat (2008) shows that demand can vary with inventory, using the case of an US magazine publisher. Chandon et al. (2009) demonstrate the effectiveness of instore marketing efforts by varying number and position of shelf facings with brand attention and evaluation at the point of purchase. Using an eye-tracking experiment, they found that the number of facings has a strong impact on consumer attention. Going from 4 to 8 facings increased the probability of noticing the item by 28% and the probability of reexamining it by 40%. But adding additionally 4 facings only added an extra 7% to noting and extra 19% to reexamination. They identify facing variation as the most significant effect among instore factors, even stronger than vertical and horizontal positioning and pricing.

3.4 Scientific Models for Master Category Planning

Furthermore, shelf space allocation has a significant impact on shelf replenishment costs. Broekmeulen et al. (2006) show that instore handling costs amount to 38% of operational logistical costs in the retail supply chain. They show that shelf space allocation is not aligned with the replenishment regime. About 60% of the items are temporarily underfaced, i.e., consumer demand is higher than shelf supply, thus requiring frequent restocking. Curseu et al. (2009) derive a model for estimating stacking time and instore handling costs. They identified case pack size and consumer units as the key cost drivers. Van Zelst et al. (2009) conclude for two different retailers that case pack sizes (and thus indirect facings) and the filling regime may deliver profit gains of 8–49%.

However, due to experimental complications within stores and high testing costs, experiments to date have not been sufficiently extensive to provide robust results (Abbott and Palekar 2008). Bultez et al. (1989) attribute weak results to poor experimental design, low variation in space allocations and unreliable sales data.

3.4.2.2 Optimization Models for Shelf Space Management

Profit optimization models formulate the demand rate as a function of the shelf space allocated to products. Space models cluster around the following three streams:

A. Space Allocation Models with Space-Elastic Demand

Common denominators of shelf space models are item i's demand d_i as a function of the space allocated to an item. The base demand is α_i, k_i is the number of facings allocated to item i, b_i is the item's breadth, β_i the (constant) space elasticity expressed as a power function, and k_i^{max} the maximum number of facings.

$$d_i(k_i) = \alpha_i \cdot (k_i \cdot b_i)^{\beta_i} \qquad k_i = 0, \ldots, k_i^{max} \tag{3.10}$$

For the first stream of space allocation models, a major study was published by Hansen and Heinsbroek (1979). They formulate the total category profit P as a maximization problem, with the number of facings as a decision variable. The vector $\bar{k} = (k_1, k_2, \ldots, k_i, \ldots, k_I)$ denotes the facings values of the items. The profit depends on the demand and the item unit profit p_i.

$$\text{Max! } P(\bar{k}) = \sum_{i=1}^{I} d_i(k_i) \cdot p_i \tag{3.11}$$

They further apply fixed replenishment costs, which are not dependent on the facing decision. A restriction ensures that only the available space S can be distributed: $\sum_{i=1}^{I} k_i \cdot b_i \leq S$. Further restrictions were minimum facings and integer constraints, as only full pack sizes were allowed. The proposed algorithm is based on a generalized Lagrange multiplier technique, which is generally only guaranteed to find local solutions of non-convex programs. The authors do not reflect cross-product relations as they predict problems in generating data for cross-product relations and the practical implementation of solution procedures.

Further developments in this stream are published by the following researchers: Zufryden (1986) developed a dynamic programming formulation for a problem with space elasticity and marketing variables. An interesting paper by Yang and Chen (1999) assumes linear profit within a constrained number of facings. They formulate a shelf space allocation problem with vertical and horizontal space allocation effects. Yang (2001) proposes a knapsack heuristic for the model. He found an optimal solution only for simplified versions. Lim et al. (2004) build on Yang's work by optimizing with meta-heuristics. A hierarchical Bayes model is proposed by van Nierop et al. (2006) to estimate interaction between shelf layout, marketing activities and stochastic demand. They optimize with simulated annealing. Murray et al. (2010) jointly model shelf space and pricing decisions and solve the problem with an MINLP-solver. Hansen et al. (2010) investigate meta-heuristics for decision models with facing-dependent demand and vertical and horizontal location effects.

Major criticism arose from the fact that cross-space effects are not reflected and heuristics are applied to find near-optimal solutions. The following subsection identifies shelf space models reflecting cross-product relations.

B. Space Allocation Models with Space- and Cross-Space Elastic Demand

In an influential paper, Corstjens and Doyle (1981) suggest a method for allocating shelf space to categories. They formulate a maximization problem based on individual and cross-product demand. They additionally include cross-space elasticity γ_{ji} in the demand calculation and procurement, carrying and OOS costs in the profit function. The demand function is constituted as follows:

$$d_i(\bar{k}) = \alpha_i \cdot (k_i \cdot b_i)^{\beta_i} \prod_{\substack{j=1 \\ j \neq i}}^{I} (k_j \cdot b_j)^{\gamma_{ji}} \qquad k_i = 0, \ldots, k_i^{max} \qquad (3.12)$$

The signomial geometric programming method optimizes shelf space for categories, but Borin et al. (1994) show that the solutions reported violate the constraints in seven out of ten cases. In addition, the multiplicative model predicts zero demand for a given category if the space of any other category is set to zero (i.e., $k_i = 0$). Further limitation lies in the narrow number of items, as the model is supposed to optimize product groups rather than individual items.

Bultez et al. (1989) apply Corstjens' model at brand level, assuming identical cross-elasticity within product groups. Borin et al. (1994) extend the demand function by differentiating the demand components. They describe unmodified, modified, acquired and stockout demand. Modified demand arouses from instore attractiveness of products. Acquired demand captures assortment decisions.

3.4 Scientific Models for Master Category Planning

Stockout demand arises when total demand exceeds shelf inventory and consumers switch to other items. They assume constant market size and distribute the volume of delisted items according to the market shares of remaining items. A simulated annealing heuristic optimizes return on inventory. They consider substitution effects due to temporary or permanent unavailability of products. However, besides the exclusion of space elasticity, they also neglect operational costs. Irion et al. (2004) further extend Corstjens' model to a product level instead of category level. Demand consists of space and cross-space elasticity. They consider purchasing costs, interest rates and listing costs. The model chooses integer facings to optimize profit under shelf space and facing constraints. Using a linearization framework, they transform the model into a mixed integer problem with linear constraints. Their approach provides only near-optimal solutions with a posteriori error bound. They use a ln-function for their linearization framework, which unfortunately does not allow accounting for zero facing demand of delistings. The model requires a predefined assortment, given the latent consumer demand for delisted items. Gajjar and Adil (2008) and Gajjar and Adil (2011) build on Irion's study and develop a local search heuristic.

However, none of the model presented reflect operational constraints of shelf replenishment, which is part of the third stream:

C. Integration of Shelf Space with Inventory Problems by Modeling Replenishment and Space Allocation

Most literature on retail shelf space management focuses primarily on the demand side, and less on the cost side. However, retailers with limited space face a trade-off of putting fewer items out for sale against keeping inventories of other products. Proper control of retail costs requires balancing warehousing, transportation, inventory, shelf space and instore handling costs (Curseu et al. 2009; van Zelst et al. 2009; Kuhn and Sternbeck 2011). Ketzenberg et al. (2002) demonstrate the profit effect of replenishment and case pack sizes on store space management with a maximum profitability point by relying on substitution to reduce space requirements. Inventory systems are consequently included in space allocation in a third stream of studies.

Urban (1998) provides the first enhancement with available inventory and replenishment systems. The deterministic, continuous-review model takes into account inventory-elastic demand, since sales before replenishment reduce the number of items displayed. Consequently, the effective shelf space assigned to products diminishes until replenishment takes place. The item profit π_i comprises unit profit p_i less the fixed costs of replenishment FCR_i, inventory holding costs h_i, and listing costs LC_i, in the case without front shelf depletion. Urban formulates order quantity q_i and facings k_i as continuous decision variables.

$$\pi_i(\bar{q},\bar{k}) = d_i(\bar{q},\bar{k})\left[p_i - \frac{\text{FCR}_i}{q_i}\right] - h_i \cdot \frac{q_i}{2} - (h_i + \text{LC}_i)k_i \qquad (3.13)$$

The multiplicative demand of item d_i depends on space-, cross-space-elastic effects and substitution from delisted items to i. Urban optimizes total category profit with Max! $P = \sum_{i=1}^{I} z_i \cdot \pi_i(\bar{q},\bar{k})$. The binary variable z_i is 1 if a product is included in the assortment. The model also covers restrictions in back room capacity, minimum order quantities and ensures that replenished quantity meets demand. The problem is solved with a generalized reduced gradient algorithm (GRG) and genetic algorithm (GA). He reports that both heuristics significantly outperform a proportional space allocation rule, and the GRG yields solutions that are within 0.4% of the solution obtained by the GA. The facings are continuous values and thus violate integer requirements for the facings. They are rounded only afterwards to integer values and may result in suboptimal profit values. Further limitations are in the restriction to individual product replenishment, but retailers normally jointly replenish products because of joint delivery cycles from central warehouses.

Inventory control integration with space allocation is proposed by Hwang et al. (2005). They optimize order quantity, facings and shelf positioning using a GRG heuristic and a GA, but exclude assortment decisions. Swami et al. (1999) develop an integer model for shelf space allocation in a movie shop. Hwang and Hahn (2000) determine an optimal procurement policy for items with an inventory-level-dependent demand rate and fixed lifetime. The demand rate is assumed to be a function of current inventory level, while each item has a fixed expiry date. Abbott and Palekar (2008) determine – exactly for a single-product case and approximately for a multi-product case – the optimal replenishment cycles for products given the costs of restocking and sales effects of inventory-elastic demand. They formulate an economic order quantity problem with time-varying demand. The optimal replenishment times have an inverse relationship to initial space assignment and space elasticity. However, the model requires an initial space assignment as input. It therefore does not optimize assortment and facing. A very comprehensive study was carried out by Hariga et al. (2007). They propose an optimization model to determine assortment, replenishment, positioning and shelf space allocation under shelf and storage constraints. The decision variables are display locations, order quantities, and the number of facings in each display area. The non-linear problem could be solved exactly for a four-item case, but requires a heuristic for a larger, practical case. They also omit integer facing values. Ramaseshan et al. (2008, 2009) determine heuristic shelf space allocation, product assortment, and inventory quantities. Their model uses the GRG to generate an approximate solution for up to 14 items (Ramaseshan et al. 2008). The demand function requires $k \geq 1$. Hübner and Kuhn (2011d) extend shelf space management models via OOA substitution. They integrate demand estimates for product delistings and the effect on other products. The model is capable of dealing with large, realistic category sizes with up to 250 items using CPLEX. Furthermore, the model reflects basic supply levels ensuring appropriate service levels with limited replenishment capacity. A further development of the model takes into account instore logistics costs and inventory holding costs (Hübner and Kuhn 2011e).

Conclusions from the Review of Shelf Space Models

It has been shown that the models discussed address specific concerns of space and inventory allocation. Furthermore, our discussion demonstrates that application of the optimization models should lead to additional profit gains over the traditional, simplistic rules of proportional space allocation used in software applications, by taking into account a broader perspective on relevant demand and cost effects. Table 3.5 summarizes the major shelf space studies and structures them according to key characteristics.

However, the models also have some limitations and drawbacks, such as only near-optimal solutions by application of heuristics, only limited scope of products, and the examination of product groups rather than individual products. There are also complications in data estimates, especially with regard to cross-space elasticity (Kök et al. 2009). Additionally, the different assumptions result in varying cost and revenue integration. Not all studies take into account the operational costs of lower facings, as fewer facings need to be restocked more frequently to serve consumer demand. The third stream shows that some of the relevant costs are not actually independent of shelf space allocation.

The biggest drawback of these studies is their limited applicability for assortment decisions. The generally proposed method of handling assortment decisions and latent demand for non-listed items (i.e., facing set to zero) is to assume no demand. Thus, these models do not integrate latent consumer demand if a product is not available in the shop, but in the shoppers' mind, and the willingness to substitute is there. Urban (1998) constitutes an exception by multiplicatively integrating the substitution demand into the demand calculation. However, he omits integer facing constraints and the numerical examples do not include assortment effects. Furthermore, the publications on assortment planning calculate substitution demand additively, rather than multiplicatively as proposed by Urban (1998). The deterministic models require that shelves are always stocked and OOS does not exist.

3.5 Conclusions and Future Areas for Research

This chapter structured commercial and scientific models in key CM areas and elaborated the interrelations. In the course of the discussion it was identified that decisions on assortment and shelf management relate closely, but are not intertwined

Table 3.5 Overview of literature on shelf space management models

Authors (by year)	Sp[a]	CrSp[a]	Additional demand effects	Costs	Function[b]	Solution method	Items[c]
Hansen and Heinsbroek (1979)	X			Restocking, OOS constraints	C	Lagrange, specialized heuristic	6,443
Corstjens and Doyle (1981)	X	X		Procurement, inventory, OOS	C	Geometric programming	5
Zufryden (1986)	X		Marketing effects	Procurement, inventory, OOS	C	Dynamic programming	40
Bultez et al. (1989)	X		Promotion, visibility	Restocking	C	Specialized heuristic	4
Borin et al. (1994)	X	X	Assortment		C	Simulated annealing	18
Urban (1998)	X	X	Assortment	Order size, inventory	C	GRG, GA	54
Yang (2001)	X		Positioning		C	Knapsack algorithm	10
Irion et al. (2004)	X	X	Marketing effects	Procurement, inventory, listing	PL	MIP model (LINDO)	6
Lim et al. (2004)	X	X		Positioning, procurement, inventory	C	Tabu search, squeaky-wheel	100
Hwang et al. (2005)	X		Order quantity, positioning	Procurement, inventory, ordering, display	C	GRG, GA	4
Hariga et al. (2007)	X	X	Positioning, order quantity	Ordering, inventory	C	MIP model (CPLEX)	4
Abbott and Palekar (2008)	X	X	Shelf depletion	Restocking cycle	C	Upper bound heuristic	4
Gajjar and Adil (2008)	X		Positioning		PL	Specialized heuristic	200
Ramaseshan et al. (2008)	X	X		Procurement, shelf space, backroom	C	GRG	14
Hansen et al. (2010)	X	X	Positioning		C	Meta-heuristics, simulations	100
Murray et al. (2010)	X	X	Positioning, price		C	MINLP model (BONMIN)	100
Hübner and Kuhn (2011d), Hübner and Kuhn (2011e)	X	X	Substitution, replenishment	Listing	C	MIP model (CPLEX)	250
Hübner and Kuhn (2012)	X	X	Substitution, price	Listing	C	MIP model (CPLEX)	250

[a] *Sp* space elasticity; *CrSp* cross-space elasticity
[b] *C* continuous, *PL* piecewise linear
[c] Maximum number of items considered in test case

3.5 Conclusions and Future Areas for Research

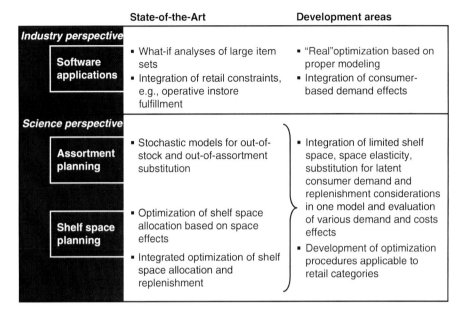

Fig. 3.4 Summary: State-of-the-art and future areas for research

in software applications and scientific models. Additionally, both retailing research and practice are not sufficiently integrated. Figure 3.4 summarizes the state-of-the-art and the development areas.

3.5.1 Alignment of Software Applications and Science

Practitioners tend to use simplistic software tools that can handle large-scale assortments. Science and practice reveal a large discrepancy in this regard. Academic models have greater analytic capabilities than available commercial software, yet they still need to prove their implementation ability. Science focuses on the approach of a large-scale integration of extensive interdependencies of demand, resulting in complicated and expensive estimation requirements for parameters. However, a major drawback is often the limited number of items that can be modeled. On the contrary, retailers adhere to more strictly functional constraints in their operational planning. Additionally, Borin and Farris (1995) show that managers tend to be overconfident in some aspects of their decision making because they underestimate the range of possible outcomes. However, the extensive research into shelf space management shows that the practical "rules of thumb" used in the commercial applications can be improved.

3.5.2 Alignment of Assortment and Shelf Space Management

Research has developed advanced models in only loosely integrated research streams of substitution and shelf space allocation. It has been shown that a wide range of models already exists in the literature that focus on separate issues limited to one area, treating related CM decisions as constraints, primarily modeling demand-based interdependencies, and considering highly stylized problems with very few items only. The literature on space elasticity deals mainly with space effects on additional sales, but only Urban (1998) integrates substitution effects. The substitution literature mainly covers stochastic effects in the case of stockouts, but omits space limits or space elasticity effects. Yücel et al. (2009) have already proved that considering shelf space in assortment policies increases profit and allows one to build more realistic setups.

3.5.3 Alignment with Other Planning Objectives

Retailers are constrained in their shelf replenishment due limitations on the shelf merchandizers available to immediately fill the shelves after stockout and the expensive handling costs within stores (Thonemann et al. 2005; van Zelst et al. 2009). Integration of replenishment constraints and non-linear shelf stacking costs is required. Furthermore, a more comprehensive integration of consumer decision processes and retail constraints may sharpen decisions. For example, McIntyre and Miller (1999) evaluated the impact of joint pricing and assortment decisions. They conclude that both pricing and assortment influence substitution and complementary effects. Additionally, they perceive joint decisions in selection, pricing, and space allocation as area of further research. Overall, however, none of the streams deals with an integrated listing and facing problem that takes into account stochastic demand, space-elastic effects and consumer-driven substitution, replenishment costs and constraints, and reflects a planning hierarchy.

3.5.4 Alignment within Shelf Space Competition

A recent area of research is the analysis of effects in category captainship. For example, Kurtulus and Toktay (2009) and Kurtulus and Toktay (2011) model effects of retailer dominated CM vs. category captainship by manufactureres. Martín-Herrán et al. (2006) and Martínez-de Albéniz and Roels (2011) analyze the competitive dynamics of pricing and shelf space assignment among competing manufacturers at a retail outlet. A comprehensive study will need to adress all relevant negotiation subjects between manufacturers and retailers, e.g., assortment, prices and shelf space.

3.5 Conclusions and Future Areas for Research

3.5.5 *Summary*

An ideal shelf space management model proposes a planning process and optimizes assortment, space allocation and replenishment, taking into account

- Instore demand functions based on space- and cross-space elasticity and substitution effects for unavailable items.
- Individual product profitability covering price, unit costs, listing costs and restocking costs.
- Restocking and inventory constraints, and target service levels.
- Actual problem sizes of a grocery category.

Considering these areas within an integrated framework and decision logic, or even simultaneously if complexity allows, will foster: (1) discussion on solving inventory and demand interdependencies; (2) the transfer of insights between planning domains; (3) bridge the current gap between scientific modeling, practical implementation and retail constraints.

The following chapter develops such a decision model for integrated assortment and shelf space planning.

Chapter 4
Assortment and Shelf Space Planning

4.1 Introduction and Motivation

As we elaborate in Chap. 3, one of the core strategic decisions grocery retailers must take involves determining their assortment and allocating it to the shelves. Retailers need to match consumer demand with shelf supply by balancing variety (number of products) and shelf service levels (number of items of a product). Offering broader assortments thus limits the appropriate service levels and vice versa, as shelf space is scarce. Retailers and producers try to satisfy consumers' needs with the right merchandise at the right store at the right time. However, the continually increasing number of consumer goods is in conflict with the fixed and scarce resource of shelf space. Consequently, retailers need to make same-time decisions on which products to offer ("listing"), and how much space to allocate for each product ("facing"). A shelf management model needs to balance supply and demand effects, as firstly listing decisions affect possible demand substitutions from delisted to listed items. Secondly, facing decisions as space allocation impact the space-dependent demand and the frequency of refill operations. However, current space allocation models mainly focus on demand modified by space effects, whereas substitution models mainly cover out-of-assortment or out-of-stock effects, but do not consider space effects and the additional effort required for refilling depleted shelf inventory between two basic replenishment periods.

Empirical studies demonstrate the impact of space on sales (see Sect. 3.4). Chandon et al. (2009) reveals that facing variation is the most significant instore factor – even stronger than positioning and pricing – using an eye-tracking experiment. Although these studies give different estimates of space-elasticity, they all recognize the positive impact of shelf space on demand and demonstrate that the sales increase is subject to marginal decreasing return. Also, there is empirical evidence that assortments have become so excessive that reducing variety significantly increases sales (Iyengar and Lepper 2000; Sloot and Verhoef 2008). Also, Drèze et al. (1994), Iyengar and Lepper (2000) and Dhar et al. (2001) report a positive impact of assortment size reductions and item delistings on demand. Boatwright and

Nunes (2001) found that significant item reductions (up to 54%) resulted in an average sales increase of 11% across 42 categories examined, and sales growth in more than two-thirds of these categories. Consequently, Grocery Manufacturers Association et al. (2005), Griswold (2007), Christiani et al. (2009) and Campillo-Lundbeck (2009) demonstrate the value of space management initiatives if they reflect the relevant consumer demand types, instore logistics efficiencies and in-stock positions. Therefore, shelf space management has received high priority from retailers and researchers.

We propose an innovative application by integrating assortment and space allocation simultaneously in one model. It combines the classic shelf space model with consumer-driven substitution effects arising from delisted items. Second, a minimum basic supply level is applied to ensure efficient and feasible shelf inventories, while providing an adequate consumer service level. Third, the suggested modeling approach is able to solve problems related to real category sizes by transferring the mixed-integer non-linear problem into a multi-choice knapsack problem with predefined demand values. These points are addressed in the context of high velocity grocery store items. Grocery shopping has unique characteristics, mainly arising from the product attributes of fast-moving consumer goods. The permanent assortment is regularly re-purchased by shoppers and replenished by the retailer.

The remainder is organized as follows. After setting the context in Sect. 4.2 and reviewing the respective literature on shelf space and substitution models in Sect. 4.3, the model is developed in Sect. 4.4. Computational tests are presented in Sect. 4.5. The final Sect. 4.6 discusses the implication of the results on retail practice and gives an outlook on further areas of research.

4.2 Problem Definition

We develop a model for determining the assortment and allocating shelf space for a grocery category, taking into account space elasticity and substitution effects. The optimization problem can be summarized with an objective function that maximizes category profitability using merchandising variables as the decision variables, and incorporating shelf replenishment and various other constraints. We assume a retailer who needs to select items from a set of $i = 1, 2, \ldots I$ items within a category. Listed items are denoted by the set N^+ and delisted items by N^-. Thus, $N^+, N^- \in I$, $N^+ \cup N^- = I$ and $N^+ \cap N^- = \emptyset$. The total demand for a listed item i ($i \in N^+$) depends on the number of facings k for all corresponding items and substitution demand from the delisted items j ($j \in N^-$) to i. Substitution demand is determined by the substitution intensity and the latent demand for delisted items d_j ($j \in N^-$). Category profit depends on the demand realized, item profit p_i and listing costs LC_i. The decision model needs to determine the facings and listing of each item to maximize total category profit P under replenishment constraints. The binary variable z_i is set to 1, if an item is listed, otherwise to 0.

4.2 Problem Definition

$$\text{Max ! } P(\bar{z}, \bar{k}) = \sum_{i=1}^{I} \left[z_i \cdot d_i(k_i) + \sum_{\substack{j=1 \\ j \neq i}}^{I} d_j(\bar{z}, \bar{k}) \cdot \mu_{ji} \right] \cdot p_i - \sum_{i=1}^{I} z_i \cdot \text{LC}_i \quad (4.1)$$

4.2.1 Properties of Demand Function

The demand of an item i is a composite function of basic demand as an estimate for unmodified demand (α_i), space elasticity effects (β_i), substitution effects (λ_j, μ_{ji}) and lost sales from consumers unwilling to substitute (η_j). Table 4.1 summarizes the modular demand effects.

4.2.1.1 Space-Dependent Demand

Common denominators of shelf space models are item demand rates as a function of the space allocated to an item.

$$d_{ik} = \alpha_i \cdot (k \cdot b_i)^{\beta_i} \quad k = 1, \ldots, K \quad i = 1, \ldots, I \quad (4.2)$$

The demand rate d_{ik} of the item i is a deterministic function of its displayed front-row facing level k. In accordance with prior research, the space-demand relationship β_i is modeled with elasticity as a power function. The breadth of an item i is given by b_i. The set of facings is denoted by $k = 1, \ldots, K$.

The base demand α_i is calculated as $\alpha_i = d_{im}/(b_i \cdot m_i)^{\beta_i}$ to calibrate the observed demand with the observed space effect. The observed (or forecasted) demand d_{im} of item i arose at observed facing level m_i.

Single cycles are considered where prices and other marketing effects are stationary. This also approximates the situation over multiple periods with average prices and constant marketing activities. Positioning, i.e., shelf layer, block building or aisle location, has more of a qualitative character. These effects require quantifications of shopping ways within the store, consumer search and category roles (see for example in Drèze et al. 1994; Mantrala et al. 2009). Consequently, these qualitative effects are not modeled under given demand and profit assumptions.

Table 4.1 Demand types of the CASP-model

Unmodified demand	Modified demand	Acquired demand
Preference effect: α_i	Instore-support effect: β_i	Assortment effect: $\mu_{ji}, \eta_i, \lambda_i$

4.2.1.2 Cross-Space Dependent Demand

Discussion on cross-space effects is ambiguous in the pertinent literature. On the one hand, some shelf space models deal with the effects of cross-space elasticity as proposed by Corstjens and Doyle (1981). This quantifies the effects of neighboring items j on the sales of an item i. On the other hand, Brown and Lee (1996) and Kök et al. (2009) describe that there is no empirical evidence that product level demand can be modeled with cross-space elasticity. Zufryden (1986) argues that consideration of cross-elasticity at an individual level would be impossible in practice due to the overwhelming number of cross-elasticity terms that would need to be estimated. Also, the effects of cross-space elasticity measured appear to have only limited influence on sales. However, we contribute to the discussion by modeling the cross-space elasticity and analyzing the resulting effects with respect to additional gains in objective value and the modified solution structure. In this chapter we use an exogenous demand substitution matrix. It describes the demand shifts between items related to the space allocation. These effects are cannibalization and gains from other items or retailers. Using the parameter μ_{ji} reduces the data requirements from I^K for all facing combinations to a matrix with I^2 combinations.

4.2.1.3 Substitution Demand

Assortment decisions with limited space imply that potentially not all products can be listed, or it may be more profitable to list other products to force consumers to switch to more profitable substitutes. Substitution effects are hence an integrated part of the model. We follow the general idea of the literature on inventory management in using exogenous estimates. Every consumer chooses their favorite item j from set I. If their favorite product j is not available for some reason, probability μ_{ji} predicts that a consumer will choose a second favorite i, $i \in N^+$. The fraction of consumers who is willing to compromise their initial choice is expressed in the probability of $1 - \eta_j$. Similar to the studies of Smith and Agrawal (2000) and Yücel et al. (2009), one round of substitution is allowed to listed items ($i \in N^+$). If consumers want to substitute their first choice with a product that is not listed ($i \in N^-$), the sales are consequently lost. In accordance with Yücel et al. (2009, p.761), it is assumed that "if a product is not available (...), it is substituted by another product or a lost sale occurs." There is no attempt to model individual consumer decisions, instead, an exogenous model is applied capable of capturing aggregated consumer demand. This allows handling of the complexity arising from multiple product relationships. The resulting model is cruder than some other substitution models, but has the advantage of being much easier to analyze and requires less data. The corresponding substitution matrix μ_{ji} can be estimated directly with market research as proposed by Kök et al. (2009).

A shortcoming of standard space models is the assumption of "zero" demand when assigning "zero" facings for delisted items. Given that a latent consumer

4.2 Problem Definition

demand exists, i.e., consumers have products in their mind but cannot find them in the shop, the classic "zero-facing zero-demand" property omits substitution. It is assumed that the demand for delisted items ranges from zero demand to the demand for one facing. We therefore define the parameter λ_j that gives the demand at one facing which is still available at zero facings. The resulting demand is named "latent demand". This latent demand is expressed by $(\lambda_j \cdot \alpha_j)$. λ_j is approximated by $1 - \beta_j$, i.e., the higher the space elasticity, the lower the demand for delisted items.

$$d_{j0} = \lambda_j \cdot d_{j1} \qquad j = 1, \ldots, I \tag{4.3}$$

The weight λ_j is a percentage of the demand at one facing. λ_j can be assumed at $1 - \beta_j$, i.e., the higher the space elasticity, the lower the demand will be assumed at zero-facings. The following Fig. 4.1 illustrates this condition:

4.2.2 Instore Inventory Management and Shelf Replenishment

Broekmeulen et al. (2006) and Kuhn and Sternbeck (2011) show that instore handling costs amount from 38% to 48% of retail logistics costs, often because shelf space allocation is not aligned with the replenishment regime. About 60–70% of the items are temporarily underfaced, i.e., consumer demand is higher than shelf supply, thus requiring frequent restocking (Broekmeulen et al. 2006; Hübner and Kuhn 2011e). Shelf space decision models often neglect this fact.

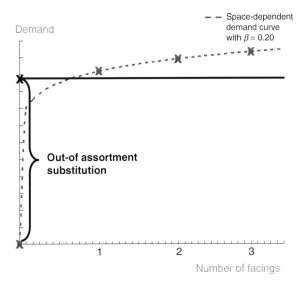

Fig. 4.1 Example for estimate of latent demand

We integrate operative replenishment requirements into the strategic model and therefore introduce a "basic supply level" (BSL), which is a percentage of demand that is covered by the basic filling process. The BSL allows operational flexibility at the shop floor level, but also limits refill frequency and quantity. Disregarding the refill frequency (i.e., without the BSL) could theoretically result in all items having a minimum number of facings and store merchandisers being permanently engaged in refilling shelves.

We prefer to follow actual retail practice by positing a joint replenishment system: all items are replenished together at regular intervals using scheduled filling with a fixed quantity, e.g., before the store opens (van Ryzin and Mahajan 1999; Smith and Agrawal 2000; Kök and Fisher 2007; Kök et al. 2009). In addition, we assume that if the shelf inventory of a product is lower than the expected demand between two regular refill periods, the sales employees have to refill the remainder with extra efforts. It therefore seems more appropriate to apply a service level constraint that ensures that the volume provided via the basic shelf filling processes will satisfy a certain proportion of consumer demand. A BSL of 100% means that total demand is fully satisfied by the scheduled basic stocking. A lower BSL requires limited individual restocking. Figure 4.2 illustrates the space-dependent demand curve, the space-dependent supply curve, and the basic supply level constraint at a level of 80% of demand.

Common retail practice is to move items forward to the front row. This allows modeling the demand for the front-row facings and not for the entire showroom inventory. The replenishment system assumed avoids temporary stockouts and partial shelf depletion. It also requires that showroom demand can be replaced immediately with backroom stock. The showroom inventory system involves a set

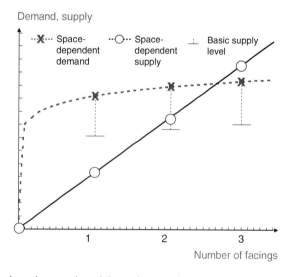

Fig. 4.2 Facing-dependent supply and demand curve without substitution and cross-space effects

of products with a display capacity S. Behind each facing are g_i units of item i that can be stored in the shelves. As fast-moving consumer goods are high-velocity items and retailers want to have high inventory levels in the showroom, interest rates are assumed not to be decision-relevant. Applying capital costs would need to cover not only showroom storage, but the entire supply chain.

4.3 Literature Review of Assortment and Shelf Space Planning

This literature review is approached from the perspective of our problem setting of integrated assortment and shelf space management, and relates to the replenishment system described. An extensive literature review is given in Chap. 3.

4.3.1 Assortment Planning Models

The literature on assortment planning models focuses mainly on defining inventory levels in the event of stochastic demand. Pentico (2008), Kök et al. (2009) and Mantrala et al. (2009) additionally provide literature reviews. Either exogenous demand or utility-based models are used to estimate consumer-driven demand substitution (Kök et al. 2009).

Exogenous demand models: Inventory models apply mainly exogenous demand estimates (see Netessine and Rudi 2003; Kök and Fisher 2007). Consumers choose from an item set I, and if the preferred item is not available for some reason, an individual consumer might accept another item as a substitute according to a predefined substitution probability. These models generally allow only one round of substitution. If the alternative product is also not available, the sales are lost (Smith and Agrawal 2000; Kök and Fisher 2007; Yücel et al. 2009). Smith and Agrawal (2000), Rajaram and Tang (2001) and Kök and Fisher (2007) analyze product substitution with heuristics based on the newsboy model. Yücel et al. (2009) study an assortment and inventory problem under consumer-driven demand substitution.

Utility-based models: In multinomial logit (MNL) models, consumer choices have recently been applied to assortment and inventory planning problems as well. Annupindi et al. (1997) model consumer arrivals in the context of vending machines. Van Ryzin and Mahajan (1999) and Mahajan and van Ryzin (2001) maximize consumer utilities based on a newsboy problem. Demand is independent of number of facings. Identical prices and costs applied for all items. Cachon et al. (2005) extend these models with consumer research costs. Identical prices are relaxed in the paper of Bish and Maddah (2004), whereas dynamic demand substitution is formulated in Hopp and Xu (2008).

MNL models are more stylized for integrating consumer choice behavior. Exogenous demand models are more flexible and allow accounting for more realistic inventory problems involving for example varying prices or pack sizes (Kök et al. 2009). However, these assortment models do not integrate space-elastic demand and shelf space limits. Only Kök and Fisher (2007) and Yücel et al. (2009) include scarce shelf space in their considerations and provide evidence of the impact of space limitations. Yücel et al. (2009) prove the significant impact of shelf space on the performance of assortments. High substitution and the costs of lost sales result in extended assortments, increased service levels and reduced profits due to increases in operational costs. Non of the models described, however, take into account space elasticity effects. Furthermore, taking into account latent consumer demand for non-available items requires a demand estimate for these items, as substitution increases demand for the listed and faced items. We thus extend our review to shelf space planning models.

4.3.2 Shelf Space Planning Models

Shelf space models formulate the demand rate as a function of the shelf space allocated to products. Hansen and Heinsbroek (1979) and Yang (2001) study shelf space allocation with demand as a function of space allocation for individual items. Corstjens and Doyle (1981) additionally include cross-space elasticity γ_{ji} in the demand calculation (see (3.12)). However, the multiplicative model predicts zero demand if one facing is set to zero. The model does not provide integer solutions as requested in such problems. Borin et al. (1994) show that the solutions reported violate the constraints in seven out of ten cases. The narrow number of items is a further limitation, as the model is supposed to optimize categories rather individual items. Borin et al. (1994) differentiate demand into unmodified, modified, acquired and stockout demand. Modified demand arises from instore attractiveness of products. Acquired demand captures assortment decisions. Stockout demand arises when the total demand exceeds the shelf inventory and consumers switch to other items. The authors assume a constant market size and distribute the volume of delisted items according to the market shares of remaining items. Besides the exclusion of space elasticity, they also neglect operational costs. Syring (2003) amplifies this with a set of logic decision rules as constraints. An interesting paper by Irion et al. (2004) extends the modeling to product level instead of entire categories. Using a linearization framework, they transform the model into a mixed-integer problem. Their approach provides near-optimal solutions with a posteriori error bound, but does not allow accounting for delistings. Gajjar and Adil (2008) build on Irion et al. (2004) and develop a local search heuristic. Ramaseshan et al. (2008, 2009) determine space allocation with a gradient search heuristic.

Finally, Urban (1998) and Hariga et al. (2007) integrate shelf space effects into inventory decisions. Both formulate the order quantity and facings as continuous decision variables, thus violating the integer constraint for the facings. Restriction to individual product replenishment represents a further limitation. However, retailers normally jointly replenish products because of joint delivery cycles from the central warehouses. The model of Hariga et al. (2007) is only capable to handle four items. Similarly, Hwang et al. (2005) optimize with a gradient search heuristic and a genetic algorithm for order quantity, facings and shelf positioning. The optimal replenishment cycles are determined exactly for a single-product case and approximately for the multi-product case by Abbott and Palekar (2008). However, Abbott and Palekar (2008) do not consider facing allocation.

Four main limitations can be identified: (1) only near-optimal solutions by using heuristics, (2) limited scope of products, (3) complicated data estimates, especially for cross-space elasticities, and (4) the fact that assortment decisions are largely disregarded. The generally proposed method for handling delisted items (i.e., facing set to zero) is to assume no demand.

To summarize, current retail shelf space literature does not simultaneously integrate space effects and substitution effects for the demand of delisted items, with respect to the actual replenishment system. The shelf space planning models deal mainly with space-dependent demand, but mostly do not integrate substitution effects for delisted items. The assortment planning models investigate substitution effects for unavailable items, but without space-depended demand and mostly without regard to limited shelf space. In the following section, we will develop such an integrated model of assortment and shelf space planning.

4.4 Formulation of the Capacitated Assortment and Shelf Space Problem (CASP)

4.4.1 Objective Function

In this section we develop a MIP model that addresses the shelf space management problem. The objective is to maximize category profit P, which is comprised of *total direct profit* (TDP), *total substitution profit* (TSP), *total cross-space profit* (TCSP) and *total costs of listing* (TCL):

$$\text{Max! } P = \text{TDP} + \text{TSP} + \text{TCSP} - \text{TCL} \qquad (4.4)$$

4.4.1.1 Total Direct Profit (TDP)

The demand function (4.2) is used to precalculate the demand d_{ik} for each possible integer facing level k of each item i. The MINLP can be degenerated into a bounded 0/1 multi-choice knapsack problem by precalculating all integer demand values, as there is a set of $i = 1, \ldots, I$ items, and each i is associated with size b_i, profit p_i and a knapsack with capacity S. The decision variable y_{ik} selects the most profitable item-facing combination from the set of available items I and facings K. The bounded knapsack model decides how many facings of each item have to be placed in the knapsack to maximize profit under capacity constraints (Pisinger 1995, 1999; Kellerer et al. 2004).

$$\text{TDP} = \sum_{i=1}^{I} \sum_{k=1}^{K} y_{ik} \cdot d_{ik} \cdot p_i \tag{4.5}$$

4.4.1.2 Total Substitution Profit (TSP)

Equation (4.6) describes the total substitution profit of item i assuming a given substitution volume from the set of delisted items, N^- to item i, $d_i^{N^-}$. However, the substitution demand can only be realized and increase i's demand, if item i is listed, i.e., $z_i = 1$. The substitution volume of item i is given by (4.7). The term $d_{j0} = d_{j1} \cdot \lambda_j$ determines the latent demand for delisted items, where d_{j1} represents the demand at one facing and λ_j the percentage of this demand that is latently. The rate μ_{ji} accounts for the substitution from product j to i.

$$\text{TSP} = \sum_{i=1}^{I} d_i^{N^-} \cdot z_i \cdot p_i \tag{4.6}$$

with

$$d_i^{N^-} = \sum_{\substack{j=1 \\ j \neq i}}^{I} d_{j0} \cdot (1 - z_j) \cdot \mu_{ji} \quad i = 1, \ldots, I \tag{4.7}$$

4.4.1.3 Total Cross-Space Profit (TCSP)

Equation (4.8) describes the substitution volume between items for changes in the facing levels (= cross-space effects). The demand $d_i^{N^+}$ given in (4.9) symbolizes the demand change resulting from changes in the facing level from the base level l to m of other listed items j, $j \in N^+$, $j \neq i$. We apply the same substitution rate μ_{ji} as for out-of-assortment substitution.

$$\text{TCSP} = \sum_{i=1}^{I} d_i^{N^+} \cdot z_i \cdot p_i \tag{4.8}$$

with

$$d_i^{N^+} = \sum_{\substack{j=1 \\ j \neq i}}^{I} \left[\sum_{m=1}^{l-1} (d_{jl} - d_{jm}) y_{jm} \cdot \mu_{ji} + \sum_{m=l+1}^{K} (d_{jl} - d_{jm}) y_{jm} \cdot \mu_{ij} \right] \tag{4.9}$$

$$i = 1, \ldots, I$$

4.4 Formulation of the Capacitated Assortment and Shelf Space Problem (CASP)

The first term in the brackets expresses the demand gains of item i through shifts from item j, where the facing level m of item j is lower than the facing level l observed. The second term expresses demand cannibalization of items j from item i when the facing m of item j is higher than the level l observed.

4.4.1.4 Total Costs of Listing (TCL)

Changes in shelf layout, point-of-sales material and slotting allowances cause fixed listing costs (LC_i).

$$\text{TCL} = \sum_{i=1}^{I} z_i \cdot LC_i \tag{4.10}$$

4.4.2 Constraints

The BSL sets demand and supply in relation to one another. The left side of constraint (4.11) covers direct demand and substituted demand of item i, while the right side denotes the supply capacity for basic fillings. The number of units supplied via basic refills per facing is described by g_i.

Only listed items ($z_i = 1$) need to achieve the BSL and to fulfill constraint (4.11). However, since d_i^{N-} and d_i^{N+} may take positive values even so item i is not listed ($z_i = 0$) it is necessary to ensure the validity of the constraint for delisted items by adding a large number on the right side, e.g., the maximum demand of item i, d_i^{max}.

$$\left[\sum_{k=1}^{K} d_{ik} \cdot y_{ik} + d_i^{N-} + d_i^{N+} \right] \cdot \text{BSL} \leq \sum_{k=1}^{K} y_{ik} \cdot k \cdot g_i + (1 - z_i) \cdot d_i^{max}$$

$$i = 1, \ldots, I \tag{4.11}$$

The shelf space constraint (4.12) limits the available space S that can be distributed among the items. Constraints (4.13) set limits to the upper (k_i^{max}) and lower (k_i^{min}) bound of facings. This depicts business restrictions such as the presentation of certain item types, enforces minimum listings for new products, or sets upper limits for certain shares of shelves. Constraint (4.14) ensures that only one facing can be selected for each item. (4.15) defines y_{ik} and z_i as a binary variable.

$$\sum_{i=1}^{I} \sum_{k=1}^{K} y_{ik} \cdot k \cdot b_i \leq S \tag{4.12}$$

$$k_i^{min} \leq \sum_{k=1}^{K} y_{ik} \cdot k \leq k_i^{max} \qquad i = 1, 2, \ldots, I \tag{4.13}$$

$$z_i + \sum_{k=1}^{K} y_{ik} = 1 \qquad i = 1, 2, \ldots, I \qquad (4.14)$$

$$y_{ik}, z_i \in \{0; 1\} \qquad i = 1, 2, \ldots, I \quad k = 1, \ldots, K \qquad (4.15)$$

4.5 Numerical Examples and Test Problems

Three test cases are applied to assess the performance of the shelf space model. The first test shows the need to integrate substitution in the space management decision in a simplified example with three products. The second test shows the impact on objective values, solution structure and calculation times with varying substitution levels, number of items and facings. Then, error bounds are provided for parameter estimates and managerial decisions using sensitivity analyses. Finally, we show test results for a large-scale problem. The optimal objective value P was found using CPLEX v11.1. All tests were run on an Intel Core 2 Duo CPU P8400 2.26 GHz processor with 4 GB RAM.

4.5.1 Illustrative Example for Impact of Substitution

The following example illustrates the need for simultaneous assortment and space optimization in retail settings. The main difference between retail and other industries is the inability to directly control substitution behavior. Retailers can only steer consumers to higher margin products through the assortment they choose to stock; consumers must then at least partially transfer their demand.

In the following illustrative example, we assume a latent consumer demand for three products with $d_1 = 13$, $d_2 = 5$, $d_3 = 4$, space elasticity of 0.2, unit profits of $p_1 = 9$, $p_2 = 7$ and $p_3 = 6$ and listing costs of $LC_i = 15$. All products have identical item sizes and only three slots in the shelf are available. Capacity per slot is 10 units.

The pure shelf space optimization without substitution results in $P = 104$, consisting of TDP $= 149^1$ and TCL $= 45$. All three items are listed with one facing ($k_i = 1$).

The integrated model works as follows. The cross-product substitution (μ_{ji}) like in Table 4.2 reflects consumer choices if products are out-of-assortment. The probabilities reflect the case that the main product $i = 1$ receives higher substitutions than the fringe products $i = 2$ and $i = 3$. η_j reflects the fact that consumers are partially not willing to compromise their initial choice. λ_j is set to 1.0 to simplify the illustrative example.

---

[1] $(10 \cdot 9 + 5 \cdot 7 + 4 \cdot 6) = 149$

4.5 Numerical Examples and Test Problems

Table 4.2 Substitution matrix of illustrative example

μ_{ji}	$i = 1$	$i = 2$	$i = 3$	η_j
$j = 1$		0.1	0.1	0.8
$j = 2$	0.6		0.1	0.3
$j = 3$	0.6	0.1		0.3

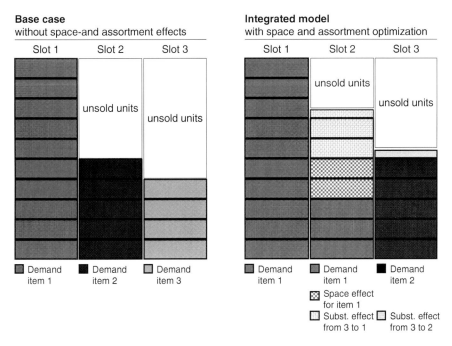

Fig. 4.3 Illustrative example for integrated assortment and shelf space planning

Figure 4.3 exemplifies the changes in shelf layout for the two models. The model lists $k_1 = 2$, $k_2 = 1$ and delists item 3 with $k_3 = 0$. The optimized profit adds up to 164 with a TDP = 169, TSP = 25 and TCL = 30. This is a 58% increase over the model without substitution. Thus item 1 profits from space effects with additional facings and cross-product substitutions, leading to a higher demand for the high-margin product. Including substitution integrates consumer decisions more accurately and impacts objective value. Joint assortment and space decisions lead to a higher shelf utilization, which increases from 63% to 76%.

4.5.2 Test Case for Hard Knapsack Problem

Five additional test examples are applied to assess the performance of the shelf space model. The first test shows that the integrated model improves objective values and that the model can be applied for practical category problems. The second analyzes the need to integrate substitution in the space management decision by the profit and solution analyses. We then show the rationale for the BSL constraint before analyzing the impact of cross-space effects on total profit and solution structure. Finally, error bounds are provided for parameter estimates using sensitivity analyses.

4.5.2.1 Description of Test Case

Data Applied in Test Case

We use demand data observed at a retail outlet within adjacent categories for the test cases. Profits are simulated to achieve correlated profit-weight combinations to test hard knapsack problems. The approach of applying such problem instances is similar to that adopted by Yang (2001) and Lim et al. (2004). The space elasticity is $\beta_i = 0.2$. An exogenous substitution estimate is applied, representing – at an aggregated consumer level – the share substituted for the first favorite in the base scenarios. The substitution intensity to the first alternative item is 0.5, to the second 0.2 and to a third 0.1. The share of lost sales is $\eta_j = 0.2$ and $\lambda_j = 0.8$. The cannibalization volume cannot exceed the demand for the items affected. We also apply bounds on the number of facings. The lower bound $k_i^{\min}$ (upper bound $k_i^{\max}$) is set 25% (400%) of the observed facings m. A BSL $= 0.8$ is applied, so that maximum of only one refilling during the period is allowed, and less than 20% of the total demand needs to be refilled during the day (Table 4.3).

The literature review shows that the scale of currently available shelf space models is restricted to a very narrow number of items. We would like to overcome this problem and allow the computing problem sizes faced by a retailer. For instance, a hypermarket carries approximately 35,000–50,000 items within 600 categories, i.e., the average number of distinct items in a category is around 60–80 (EHI Retail Institute 2010). Direct information from retailers suggests that there are no categories exceeding 200 items. One should note that the scale of our problem depends on the values of two variables: number of items (I) in the category and number of facing levels (K). Problem cases are chosen that include very high values

Table 4.3 Data for assortment and shelf space test problem

Parameter	Description, values
d_{il}	Annual demand observed, with average demand of 10,348 units and standard deviation of 10,390 units
LC_i	1,000
p_i	Uniformly distributed between $0.64 \leq p_i \leq 2.27$
$\frac{p_i}{b_i}$	Knapsack profit-to-weight correlation with $R^2 = 0.8055$, see Fig. 4.4

4.5 Numerical Examples and Test Problems

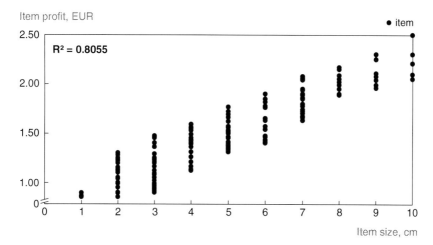

Fig. 4.4 Profit-space correlation for assortment and shelf space test problem

for the variables to ensure that the model is capable to solving realistic problem sizes for the problem being considered (Fig. 4.4).

Comparison of the CASP Solution with Other Approaches

The following definition and models are used for the numerical examples:

- BM: Base model with proportional space allocation related to market shares as commonly applied in shelf space software and substitution effects calculated a posteriori. BM_{CR} also integrates a posteriori cross-space effects.
- CSP: Capacitated Shelf space Problem with the objective function $P =$ TDP − TCL and substitution effects calculated a posteriori. CSP_{CR} additionally integrates TCSP in the objective function.
- CASP: Capacitated Assortment and Shelf space Problem with integrated space and substitution effects ($P = $ TDP+TSP−TCL). $CASP_{CR}$ also integrates TCSP.

4.5.2.2 Profit Impact of Integrated Assortment and Shelf Space Planning

Impact of Integrated Assortment and Shelf Space Planning

Table 4.4 presents the impact of integrated modeling of assortment and shelf space planning. We compare the integrated solutions of CASP and $CASP_{CR}$ with BM, BM_{CR}, CSP and CSP_{CR} for the varying number of items and facings.

All optimization models significantly outperform the BM models by a 5–10% higher profit. Integrating the substitution effects increases the objective value across all test cases. The integrated models (CASP and $CASP_{CR}$) also achieve 1–4% higher

Table 4.4 Profit impact and run time of integrated models with varying items and facings

		CASP			CASP_{CR}		
I	K	ΔP to BM (%)	ΔP to CSP (%)	Run time (s)	ΔP to BM_{CR} (%)	ΔP to CSP_{CR} (%)	Run time (s)
10	10	6.2	1.1	<1	6.2	3.7	<1
30	10	5.5	0.8	<1	4.5	0.8	1
50	10	6.0	1.6	7	4.7	1.3	97
100	10	7.8	0.8	447	5.7	0.7	531
200	10	8.4	1.6	678	9.7	1.8	1,446
250	10	8.3	1.7	747	10.0	1.6	1,534
80	5	6.3	1.4	195	4.2	1.1	708
80	10	6.2	1.0	345	5.5	1.0	641
80	20	6.6	1.0	367	5.1	0.7	637

profits than the pure shelf space models (CSP and CSP_{CR}) due to the directly integrated substitution. Transformation into a knapsack problem makes it possible to calculate optimal solutions time-efficiently for the problem being considered. The extended model is capable of handling even a large number of items and all relevant category problem sizes with very fast computation time.

Impact of Integrated Assortment and Shelf Space Planning Over Industry Practice

In the next test we compare the performance of models with BM with varying substitution intensity. All optimization models significantly outperform the BM by a 5–10% higher profit. The left side of Fig. 4.5 shows the changes in the total profit deriving from increasing substitution intensity in a test with 50 items and up to 10 facings. It reveals that the integrated models (CASP and CASP_{CR}) also achieve up to 1.7% higher profits than the pure shelf space models (CSP and CSP_{CR}). In the case without cross-space effects, the increasing substitution intensity compensates for suboptimally allocated facing and assortment decisions. In the case with cross-space effects, disregarding TCSP in the BM_{CR} decision logic erodes the profit base due to high cannibalization, leading to higher profit gaps compared to CASP_{CR} and CSP_{CR}.

The optimization models not only result in higher profits but also significantly change the structure of the result, i.e., characteristics of the decision variables. The optimization models allocate significantly different facing levels to the items. The right side of Fig. 4.5 shows that up to two-thirds of the items may be non-optimally allocated by the BM. Furthermore, the integrated assortment decision results in dependence on the substitution level in up to 40% different facing values for the items compared to the pure space models.

4.5 Numerical Examples and Test Problems

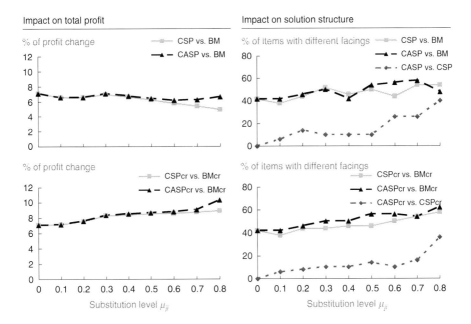

Fig. 4.5 Impact of integrated assortment and shelf space planning

4.5.2.3 Further Managerial Insights

Impact of the Basic Supply Level

The BSL ensures that total demand can be fulfilled with the inventory available. Not reflecting the BSL constraint results in high restocking requirements. Figure 4.6 shows the trade-off decision for applying a BSL, with BSL = 100%. Disregarding the BSL increases the profit by 1–8%. However, between 67% and 87% of the items receive to less facings. That means that without the BSL, not the entire demand for more than two-thirds of the items can be fulfilled with the inventory available, and therefore frequent restocking activities are necessary. Thus, a retailer needs to carefully plan the replenishment frequency and BSL.

Impact of Cross-Space Effects

If there are cross-space effects, they need to be integrated into the decision calculus. Not integrating cross-space effects (TCSP) into the objective function of the decision models would result in up to 2% lower profits. The facing allocation would be more erroneous. In this case up to 44% of items receive inappropriate facing levels (see Fig. 4.7).

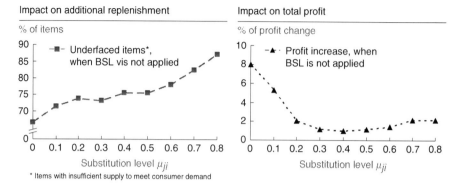

Fig. 4.6 Impact of BSL constraint

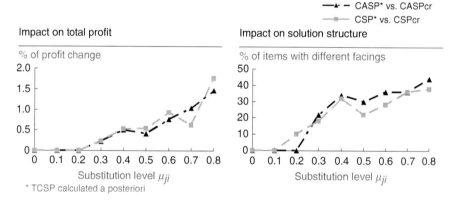

Fig. 4.7 Impact of cross-space effects: CASP$_{CR}$ vs. CASP

4.5.2.4 Sensitivity Analyses of CASP

A multitude of parameters need to be estimated. Errors and deviations cannot easily be excluded, therefore this subsection evaluates the impact of these parameters on the results of the model using sensitivity analyses. The parameters for space elasticity (β_i) and substitution levels (μ_{ji}) are estimated through consumer experiments and thus by their nature may deviate from real consumer behavior. The basic supply level and total shelf space are managerial decisions determined by different overarching parameters like store sizes, marketing activities, and category plans. Figure 4.8 shows the analyses with varying substitution levels and space elasticity. It shows moderate impact on both the decision structure and profit sensibilities. Profit sensitivities are between -1.1% and $+3.0\%$ compared to the base value. Up to 30% of the decision variables may change. The higher accuracy in estimating consumer behavior requires expensive consumer research. These sensitivity results provide business with boundaries to consider investing in understanding more accurately

4.5 Numerical Examples and Test Problems

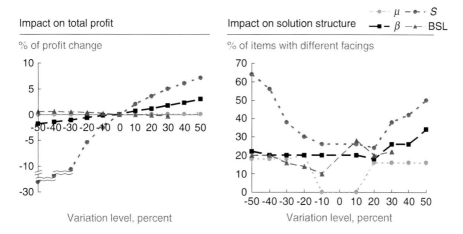

Fig. 4.8 Sensitivity analyses of the CASP$_{CR}$-model

consumers. Managerial planning decisions regarding S impact significantly the solution structure and profit prospects. Shelf space has a high downside risk, if less space has been allocated. The level of the BSL constraint has a limited effect on the total profit level, but on the space decision. It shows that a comprehensive planning framework and hierarchy would facilitate the accuracy of decision making.

4.5.3 Applicability of CASP for Large-Scale Problems

As computation time depends on facing and items, the following numerical example analyzes the computation time for a simulated test case with 10,000 items. The objective of this subsection is to analyze the computation times for a larger data sets such as an entire store. For simplicity's sake substitution is only allowed according to market shares, as a direct exogenous estimate of the substitution item by item appears difficult for larger data sets. Hence estimation of substitution volume proceeds as follows:

An alternative way to estimate substitution, especially for larger item sets is to apply probabilities with proportional market shares (see Smith and Agrawal (2000) and Irion et al. (2004)). It redistributes additional demand in proportion to the item's market shares. The proportional substitution matrix has properties that are consistent with what would happen in a utility-based framework such as the MNL model: $\mu_{ji} \geq \mu_{jl}$, if $d_i \geq d_l$ (see proof in Kök and Fisher 2007). d_i describes the market share of product i and the denominator is the remaining share.

$$\mu_{ji} = (1 - \eta_j) \cdot \left(\frac{d_i}{\sum_{\substack{l=1 \\ l \neq i}}^{N} d_l} \right) \tag{4.16}$$

Table 4.5 Evaluation of computation performance of the CASP-model

Lost sales η_j	ΔP as % of BM	Calc. time (s)
0.2	6.9%	15
0.3	6.8%	11
0.4	6.7%	14
0.5	6.5%	21

Equation (4.17) represents the substitution profit for the market share logic, whereas a substitution matrix is more suited for smaller cases with fewer items, which still allows estimating product-by-product the substitution probabilities, (4.17) fits for larger item sets. MS_j is the market share of the receiving item i, and consequently the share of of substitution from item j to i. Other data are distributed as in previous tests. Cross-space elasticity is not taken into account.

$$\text{TSP} = \sum_{i=1}^{N} \sum_{\substack{j=1 \\ j \neq i}}^{N} d_{j0} \cdot (1 - z_j) \cdot \mu_{ji}(\text{MS}_i) \cdot p_i \cdot z_i \quad (4.17)$$

The objective value increased with substitution by 7%. The computation times show that the extended model is capable of handling even large number of items like entire store settings. Times are much lower where an endogenously estimated substitution matrix is used. Thus, fewer cross-product references accelerate the computation time here even when optimizing for entire stores with 10,000 items. The computation times for a category case remains within reasonable time bounds for the strategic problem (Table 4.5).

4.6 Conclusions and Future Areas for Research

The model described extends shelf space models by substitution effects and replenishment constraints. It is based on consumer decisions affected by space effects and substitution. The operational constraints ensure that shelf space decisions are aligned with the retailer's replenishment policy. Additionally, this approach provides results for retail-specific category problem sizes within reasonable computation times. We also provide a realizable approach for the modeling of cross-space effects.

Overall, integrating substitution in the objective function not only improves the accuracy of the model, but also allows better objective values to be achieved. Standard retail data have been mainly used, such as observed sales, which are available at a store level and have been supported by experimental data. The model has been tested with correlated profit-to-weight data to test hard knapsack problems. Numerical examples prove that the model has the ability to compute all practical cases, as realistic grocery category sizes do not exceed 200 items.

4.6 Conclusions and Future Areas for Research

Limitations of the model provide input for further areas of research: (1) supply assumptions (2) demand generating effects, and (3) certainty of demand data. (1) The model assumes fixed restocking costs. These costs could be specified in more detail, i.e., quantified by the optimal restocking cycles that would result from an integrated supply model. Furthermore, retailers may also distinguish between group and individual product replenishment. Out-of-stock substitution does not occur as efficient instore logistics are assumed. However, adding restocking process costs would lead to a situation where it might be more beneficial to save on stocking, instead allowing substitution of other products. The model assumes unlimited transportation, warehouse and backroom capacity, as only showroom effects are considered. Additionally, aspects of a planning hierarchy as input to managerial constraints and enlarging the scope up the supply chain will gain additional insights. (2) Marketing activities and demand-generating effects should be also investigated. These particularly include positioning effects to account for different shelf layers and "eye-level" demand, price effects with price and cross-price elasticity, and other marketing variables that generate instore demand. (3) Finally, consumer demand is assumed to be deterministically known, but is subject to certain volatility depending on external factors like season, temperature or weekday. Assuming stochastic demand would add an important and realistic modeling feature.

In the following chapter, we address additionally the inventory and replenishment costs for assortment and shelf space planning.

Chapter 5
Assortment, Shelf Space and Inventory Planning

5.1 Introduction and Motivation

Retailers need to cost-efficiently manage the complexity of satisfying consumer demand with shelf inventory by determining the interdependent problems of assortment size, shelf space assignment and shelf replenishment. Retail shelf space assignment has demand and inventory effects, as already Whitin (1957) noted: "For retail stores (...) an increase in inventory may bring about increased sales of some items. On the other hand, an increased inventory might lead to a decrease in sales of other items." Hence, the more space and inventory is assigned to products, the higher the demand (= space-elastic demand). Furthermore, retailers need to solve the trade-off decision of either holding too much inventory, resulting in tied capital, or keeping an inventory that is too low, requiring frequent restocking. Retailers are constrained in shelf replenishment due limitations on the shelf merchandizers available to immediately fill the shelves after stock out and the expensive handling costs within stores (van Zelst et al. 2009). Proper control of retail costs requires the balancing of inventory, shelf space and instore handling costs (Curseu et al. 2009), especially as up to 48% of total logistics costs accrue in stores (Ketzenberg et al. 2002; Thonemann et al. 2005; Broekmeulen et al. 2006; Sternbeck and Kuhn 2010). For example, van Zelst et al. (2009) conclude that case pack sizes and the filling regime may deliver profit gains of 8 to 49%. While there is evidence of consumer response and cost implications, "less is known about how to translate this evidence into profitable strategies" (Kopalle et al. 2009, p.62). As we will see in the literature review, traditional shelf space management models focus on the space assignment and assume "efficient inventory management systems." In other words, they decouple the decision of shelf space assignment from replenishment, as most optimization models focus primarily on the demand side and less on the cost side. In addition, they assume that each product is restocked instantaneously and individually. Common practice in retail, however, is to conduct two types of shelf refilling:

- *Scheduled basic group filling* of products jointly by dedicated merchandizers, e.g., daily before the store opens. This is limited to a fixed weekly schedule.
- *Individual concurrent refilling* of single products by sales staff, e.g., during their idle time. Sales staff concurrently monitor shelf inventory, and instantly refill when inventory drops below a minimum level.

These different refilling processes also have cost implications. Scheduled basic group filling is more cost efficient as low-cost merchandizers jointly restock products, whereas the individual concurrent refilling is operated individually for each product, and by the more expensive sales staff. Current shelf management models, however, do not differentiate between these two types of shelf filling processes.

A further complication is that most retailers mainly plan shelf space from a sales perspective, instead from an integrated sales and logistics perspective. Retailers usually adopt commercial shelf space planning programs for creating their planograms. These tools visualize shelf space arrangement, and report product sales and profit for example. In the past, actual decisions were made by simplistic allocation rules like proportional-to-sales and a limited number of manual adjustments. Advances in computing resources should now allow the development of more complex shelf space management models that are more consistent with consumer instore behavior and required retailer planning aspects in instore logistics. Category managers from both retailers and consumer goods producers can use shelf space models to improve their decision making.

The goal of this chapter is therefore to propose an extension of shelf space management models by including restocking aspects in decision calculus. It follows the guidance of Hadley and Whitin (1963, p.21) that, "the purpose in constructing a mathematical model of an inventory system is to use it as an aid in developing a suitable operating doctrine for the system." The optimization model ensures efficient and feasible shelf inventory, clarifies restocking requirements, and allows the resolution of retail-specific problem sizes. It captures the critical decision trade-offs faced by retailers in aligning marketing-related demand effects and instore logistics-related cost effects. The planning problem will be structured, a decision support model will be provided to maximize category profit taking into account replenishment requirements, and the capability of this model will be tested to solve a category-specific problem. Specifically, this chapter extends the CASP by integrating restocking and inventory holding costs.

The remainder is organized as follows. The first Sect. 5.2, establishes the context of the decision problem and Sect. 5.3 provides a literature review. Section 5.4 develops the model, including the inventory management implications, the objective function, and alignment with hierarchical planning. Section 5.5 provides numerical examples. The final Sect. 5.6, concludes the chapter and the main results and potential extensions to the model will be discussed.

5.2 Problem Definition

This section develops a model for determining the assortment and allocating shelf space within a fast-moving retail category, taking into account replenishment cost implications. The optimization problem can be summarized with an objective function that maximizes category profitability using merchandising variables as the decision variables, and incorporating various constraints. Consider a retailer that needs to select items from a set of $i = 1, 2, \ldots I$ products within a category, and where demand d_i depends on number of facings k_i and the substitution demand μ_{ji} from item j to i for delisted items. Furthermore, the category profit depends on item profit p_i, costs of refilling, and overstocked inventory. $q_i^{(u)}$ represents demand that needs to be refilled during the day as overall demand exceeds basic shelf supply, and RFC$_i$ stands for the associated refilling costs of item i. $q_i^{(o)}$ represents overstocked inventory, c_i the unit purchase costs, and h the interest rate.

The category manager needs to decide about the facings from the demand and replenishment perspective of each item to maximize the category profit P.

$$\text{Max! } P(\bar{z}, \bar{k}) = \sum_{i=1}^{I} \left[d_i(k_i) + \sum_{\substack{j=1 \\ j \neq i}}^{I} \mu_{ji} \cdot (1 - z_j) \right] \cdot p_i - q_i^{(u)}(k_i) \cdot \text{RFC}_i - q_i^{(o)}(k_i) \cdot c_i \cdot h$$

(5.1)

Demand takes into account β_i space elasticity and μ_{ji} substitution effects for delisted items. Other marketing effects such as shelf positioning and prices are assumed to be constant. The basic constraints are: store shelf capacity, product availability, lower and upper bounds on facings. Consistent with prior research, immediate replenishment is applied to avoid temporary stock outs.

5.2.1 Inventory Management Systems and Cost Implications

This subsection completes the assumptions of the inventory management system. The showroom inventory system involves a set of products with a display capacity S. The usual retailer practice is to move items forward to the front row. This results in a model with demand depending on facings. It is assumed that retailers employ an effective logistics system that ensures that demand required in the showroom can be replaced immediately with stock available from the backroom, i.e., shortages are not allowed (Hansen and Heinsbroek 1979; Corstjens and Doyle 1981; Urban 1998; Hariga et al. 2007; Abbott and Palekar 2008; Hansen et al. 2010; Murray et al. 2010).

Two types of refilling are applied, the scheduled basic group filling and the individual concurrent refilling. Figure 5.1 illustrates the space-dependent demand and supply curves for one item. Note that supply is calculated as $(k_i \cdot g_i)$, with g_i being the capacity in units behind one facing, e.g., it can be an entire case pack. The example below shows that for $k = 1$, refilling is required three times, as

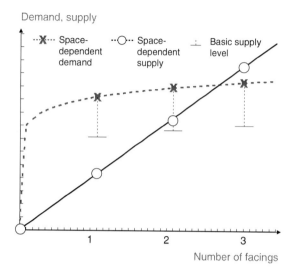

Fig. 5.1 Comparison of space-dependent demand and supply

the maximum facing-shelf capacity is less than 50% of demand at this facing. As only limited sales staff time can be allocated to shelf filling, the refilling process needs to be limited. The application of the basic supply level (BSL) constraint restricts solutions to more cost-efficient supply-demand relations, as supply points are only allowed that are above a certain BSL to avoid situations, where items are permanently underfaced (see Broekmeulen et al. 2006). The BSL needs to be achieved by the regular shelf filling process, whereas sales staff needs to fulfill only the residuum during the day. This avoids high restocking requirements between the scheduled basic filling cycles. The entire demand can be fulfilled by items available on the shelves, i.e., by basic filling (i.e., BSL) and partial refilling (i.e., 1 − BSL).

Furthermore, shelf space management with respect to replenishment requirements reflects the costs associated with undersupplied volume or overstocked inventory. If supply is lower than demand, as for $k = 2$ in the example, additional individual refilling of single products by sales staff needs to be applied. If supply is higher than demand, as for $k = 3$ in the example, then part of the inventory always remains on the shelves, and resulting in permanently invested capital.

Figure 5.2 illustrates the development of shelf inventory levels and the associated refilling processes according to the relationship between supply and demand. Accordingly, either demand exceeds supply or vice versa. If demand exceeds supply then the insufficient basic supply requires refilling between periods, or – if the contrary applies – overstocked items increase capital employed. The third possibility, where demand matches the basic supply $q_i^{(b)}$ exactly, is only theoretically possible, as supply is based on entire shelf slots.

Basic restocking costs do not depend on the number of facings. They accrue for all items during the basic group filling process. Individual refilling of faster-selling

5.3 Literature Review of Shelf Space and Inventory Planning

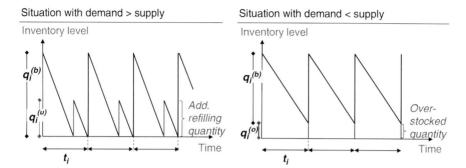

Fig. 5.2 Development of retail shelf inventory levels

items is carried out by the sales staff during the day, i.e., when demand is higher than basic supply. While basic group filling costs are not dependent on facings, and as a result are not decision-relevant for the problem, individual refilling requirements are variable costs that depend on the facings.

Interest rates are not applied on products sold between restocking cycles. This would be against retailers' overall target of having full shelves as an image factor to attract customers. Hence, retail showroom shelves are not designed to be "lean warehouses." The entire supply chain rather than only the showroom inventory would need to be reflected to generate inventory holding cost reductions, i.e., warehouses, distribution and stores. As long as the model assumes that sufficient inventory for immediate shelf replenishment needs to be located in the backroom, it does not matter from a capital investment perspective whether inventory is located in the backroom or showroom. Furthermore, fast-moving consumer goods as high velocity items have very limited storage duration in the showroom.

5.2.2 Properties of Demand Function

The demand is formulated as in Chap. 4 and (4.2). Hence, demand for an item i is a composite function of basic demand α_i as an estimate for unmodified sales, β_i for space elasticity, μ_{ji} for substitutions from j to i, and η_j as the share of consumers not willing to substitute. Table 4.1 summarizes the modular demand effects. Note that here we do not model cross-space effects as we focus on the analysis of replenishment effects.

5.3 Literature Review of Shelf Space and Inventory Planning

The planning areas considered are assortment and shelf space planning. *Assortment planning* is the listing decision based on consumer-choice behavior and substitution effects. *Shelf space planning* is the facing and replenishment decision based on space elasticity effects, limited shelf space and operational restocking requirements. This section will show that demand-side-related shelf space management models treat supply-side-related considerations as exogenous parameters. The models further assume predefined assortments, with the exception of Urban (1998) and the CASP-model in Chap. 4, as all the others omit substitution effects for non-available items.

A first major study was published by Hansen and Heinsbroek (1979). They formulate the profit P as a maximization problem, with the number of facings as the decision variable under the constraint of limited shelf space. The profit includes facing-dependent demand and item profit p_i (Max! $P(k_i) = \sum_{i=1}^{I} d_i(k_i) \cdot p_i$). The authors use a Lagrange multiplier to solve the problem, but do not reflect on cross-product relations. They further apply fixed replenishment costs, which are not dependent on the facing decision. Further, Zufryden (1986) and Yang (2001) study shelf space allocation with demand as a function of individual items' space allocation. In a second stream of model types, Corstjens and Doyle (1981), Borin et al. (1994), Bultez et al. (1989), Lim et al. (2004), Irion et al. (2004), Hansen et al. (2010) and Murray et al. (2010) develop space allocation models with demand as a function of category and individual item space allocation.

One drawback of the existing models is that they give no explicit consideration to inventory-related decisions. Some include an operating cost factor that assumes that costs are proportional to the sales of the product (Urban 1998). However, these costs are independent of inventory levels and restocking frequency. The only operational constraints these studies take into account are limited shelf space and limits on the number of facings.

However, retailers with limited space face a trade-off of either reducing assortment size and increasing inventory levels of remaining products, or increasing assortment size and reducing the number of facings, which may require more frequent restocking. Consequently, in a third stream of studies, replenishment considerations are included in space allocation models.

Urban (1998) provides the first enhancement with inventory and replenishment considerations. He takes into account inventory-elastic demand, since sales before replenishment reduce the number of items displayed. Consequently, the effective shelf space assigned to products diminishes until replenishment takes place. The profit maximization model comprises unit profit less fixed order cost FOC_i, inventory holding cost, with h as interest rates and c_i unit costs, and refilling costs RFC_i. T is the period, the order quantity q_i is the decision variable, and $\widetilde{q}$ is the average inventory in the backroom and showroom, including shelf depletion with an inverse space elasticity.

$$\text{Max!} \ P(\bar{q}, \bar{k}) = \sum_{i=1}^{I} \frac{p_i \cdot q_i}{T} - \frac{\text{FOC}_i}{T} - \widetilde{q_i} \cdot c_i \cdot h - \text{RFC}_i \cdot k_i \quad (5.2)$$

This model is extended to a multi-product assortment and shelf space allocation problem. The model also covers restrictions in backroom capacity and minimum order quantity. Urban solves the problem with a genetic algorithm. However,

5.3 Literature Review of Shelf Space and Inventory Planning

the performance of the heuristic has only been compared to proportional space allocation rules, and the model does not allow group replenishment of items, as only instantaneous individual restocking is assumed.

Hariga et al. (2007) determine assortment, replenishment, positioning and shelf space allocation under shelf and storage constraints for a limited problem size, but omit substitution effects. The decision variables are display locations, order quantity, and the number of facings in each display area. The complex mixed-integer non-linear problem could be solved exactly for a four-item case, but according to the authors would require a heuristic for a larger, practical case. Furthermore, they omit integer facing values.

Abbott and Palekar (2008) formulate an economic order quantity problem, exactly a single-product case, and approximately for a multi-product case, determining the optimal replenishment cycles for products given the costs of restocking and the sales effects of inventory-elastic demand. The optimal replenishment time has an inverse relationship to initial space assignment and space elasticity. However, it requires an initial space assignment as input. Thus, it does not optimize assortment and facing.

Yücel et al. (2009) analyze an assortment and inventory problem under consumer-driven demand substitution. They conclude that neglecting consumer substitution and space limitations has a significant impact on the efficiency of assortments. However, they do not include space elasticity effects.

The CASP-model in Chap. 4 complements a shelf space management model with out-of-assortment substitution. We integrate demand estimates for product delistings and the effect on other products. It has been demonstrated that the model is capable of dealing with large, realistic category sizes. Furthermore, the model reflects basic supply levels, ensuring appropriate service levels with limited replenishment capacity.

A detailed discussion of shelf space models can be found in Chap. 3. Also, further literature on inventory-level-dependent demand problems is provided by Urban (2005). He distinguishes models in which demand rate is a function of the initial inventory level, and models in which it is dependent on the instantaneous inventory level. The first is closer to space allocation problems, while the second relates to newsboy and assortment problems. Table 5.1 summarizes the main differences in the models reviewed related to shelf space and inventory problems.

Broekmeulen et al. (2006) report that shelf space allocation is often not aligned with the replenishment regime. About 60% of the items are temporarily underfaced, i.e., consumer demand is higher than shelf supply, thus requiring frequent restocking. Shelf space decision models often neglect this fact by assuming efficient replenishment systems and disregard instore logistics. Traditionally, space allocation has more a strategic character, whereas replenishment decisions treated as operational problems. So far, shelf space models abstract from retail requirements in their replenishment assumptions.

Usual practice of retailers is to conduct joint replenishment of products, e.g., in the morning before the store opens (Smith and Agrawal 2000; Kök and Fisher 2007). With item-specific demand rates, joint replenishment implies stock outs

Table 5.1 Retail shelf space management models related to inventory management

Criteria	Urban (1998)	Hariga et al. (2007)	Abbott et al (2008)	Yücel et al. (2009)	CASP-model Chap. 4
Inventory location[a]	B, S	B, S	S	S	S
Substitution	X			X	X
Shelf depletion	(X)	X	X		
Decision variable					
Facing	X	X		X	
Restocking[b]	q		r	q	
Assortment	X			X	X
Other		location		supplier	
Individual restocking	X	X	X	X	X

[a] B backroom, S showroom
[b] q restocking quantity, r restocking point
X: fully integrated; (X): partially integrated

between replenishment cycles if some items are sold faster than stocked. To avoid lost sales, sales employees also refill any shelf gaps that arise between the scheduled basic filling cycles. The above-noted models assume efficient restocking and omit restocking capacity constraints. They apply instantaneous and individual restocking, i.e., as soon as an item runs out of stock, retailers immediately refill the empty shelf. In particular, these models do not differentiate between scheduled basic group filling processes and individual concurrent refilling during the day, although basic shelf filling is completed by more cost-efficient merchandising systems, and refilling by the sales workforce. The sales staff refilling is more expensive as it (1) accrues for individual products and not efficiently in product groups, and (2) it is conducted by more expensive sales staff.

In this research, the following three important elements are integrated into the basic shelf space decision model: (1) assortment decisions with substitution effects for delisted items; (2) basic supply level constraint to limit number of refills and (3) introduction of group refilling processes and differentiating the associated restocking costs of individual and group refilling.

5.4 Formulation of the Capacitated Assortment, Shelf Space and Replenishment Problem (CASRP)

5.4.1 Objective Function

Objective is to maximize product category profit. The total profit P consists of TDP (*total direct profit*), TSP (*total substitution profit*), TCL (*total costs of listing*), TCUS (*total costs of undersupply*) and TCOI (*total costs of overstocked inventory*).

$$\text{Max!} \ P = \text{TDP} + \text{TSP} - \text{TCL} - \text{TCUS} - \text{TCOI} \quad (5.3)$$

5.4.1.1 Total Direct Profit (TDP)

The *total direct profit* covers the profit of items regardless of their relation to the remaining assortment. The demand (4.2) is used to precalculate the demand d_{ik} for each item i and its associated facing level k. This preprocessing enables transfer of the non-linear demand function into a 0/1 multi-choice knapsack problem where the binary variable y_{ik} selects the most profitable item-facing combination for a knapsack of the shelf size S, the item size b_i, and p_i as the item's gross profit.

$$\text{TDP} = \sum_{i=1}^{I} \sum_{k=1}^{K} y_{ik} \cdot d_{ik} \cdot p_i \tag{5.4}$$

5.4.1.2 Total Substitution Profit (TSP)

The *total substitution profit* integrates substitution profit through demand shifts for delisted items. The term $(\lambda_j \cdot d_{j1})$ symbols the latent demand for delisted items, with λ_j expressing a share and d_{j1} the demand at one facing. The substitution matrix μ_{ji} integrates transition probabilities between items j and i. The binary variable z_i indicates whether an item is listed (set to 1) or delisted (set to 0).

$$\text{TSP} = \sum_{i=1}^{I} d_i^{N^-} \cdot z_i \cdot p_i \tag{5.5}$$

with

$$d_i^{N^-} = \sum_{\substack{j=1 \\ j \neq i}}^{I} \lambda_j \cdot d_{j1} \cdot (1-z_j) \cdot \mu_{ji} \quad i=1,\ldots,I \tag{5.6}$$

5.4.1.3 Total Costs of Listing (TCL)

The *total costs of listing* are items' fixed listing costs LC_i for advertising, layout changes or slotting allowances.

$$\text{TCL} = \sum_{i=1}^{I} z_i \cdot LC_i \tag{5.7}$$

5.4.1.4 Total Costs of Undersupply (TCUS)

The *total costs of undersupply* integrate the additional refilling requirements if demand is higher than supply, expressed by the extra refilling volume $q_i^{(u)}$. The difference between supply and demand is $q_i^{(u)}$, where demand exceeds supply. RFC$_i$ depicts item-specific refilling costs, that can for example depend on the instore transportation processes from the backroom to the shelf and the refilling volume behind one facing.

$$\text{TCUS} = \sum_{i=1}^{I} q_i^{(u)} \cdot \text{RFC}_i \qquad (5.8)$$

5.4.1.5 Total Costs of Overstocked Inventory (TCOI)

The *total costs of overstocked inventory* comprise capital costs of overstocked volume $q_i^{(o)}$, i.e., where supply exceeds demand before the next basic scheduled shelf filling process. The parameters c_i are the product costs and h is the interest rate.

$$\text{TCOI} = \sum_{i=1}^{I} q_i^{(o)} \cdot c_i \cdot h \qquad (5.9)$$

5.4.2 Constraints

The constraints are composed of hierarchical planning aspects and modeling requirements. The first set of constraints (5.10–5.11) reflect the input from overarching decisions like instore layout or category planning. (5.10) ensures that only the available space S can be distributed. (5.11) sets boundaries on the number of facings (k_i^{min}, k_i^{max}), and for example enforces minimum listings. (5.12) ensures that the scheduled basic refilling frequency RF is sufficient to meet a defined basic supply level (BSL) and only the remaining demand needs to be refilled during the cycles. (5.13) and (5.14) defines the volume either for under- or oversupplied volumes. (5.15) allows only one facing level for each item.

$$\sum_{i=1}^{I} \sum_{k=1}^{K} y_{ik} \cdot k \cdot b_i \leq S \qquad (5.10)$$

$$k_i^{min} \leq \sum_{k=1}^{K} y_{ik} \cdot k \leq k_i^{max} \qquad i = 1, 2, \ldots, I \qquad (5.11)$$

$$\left(\sum_{k=1}^{K} y_{ik} \cdot d_{ik} + d_i^{N-} \right) \cdot \text{BSL} \leq \sum_{k=1}^{K} y_{ik} \cdot k \cdot RF \cdot g_i + (1-z_i) \cdot d_i^{max} \qquad i = 1, 2, \ldots, I \qquad (5.12)$$

$$q_i^{(u)} \geq \sum_{k=1}^{K} y_{ik} \cdot (d_{ik} - k \cdot g_i \cdot RF) + d_i^{N^-} - (1-z_i) \cdot d_i^{max} \quad i = 1, 2, \ldots, I \quad (5.13)$$

$$q_i^{(o)} \geq \sum_{k=1}^{K} y_{ik} \cdot (k \cdot g_i \cdot RF - d_{ik}) - d_i^{N^-} \quad i = 1, 2, \ldots, I \quad (5.14)$$

$$z_i + \sum_{k=1}^{K} y_{ik} = 1 \quad i = 1, 2, \ldots, I \quad (5.15)$$

$$y_{ik}, z_i \in \{0; 1\} \quad i = 1, 2 \ldots, I \quad k = 1, \ldots, K \quad (5.16)$$

$$q_i^{(o)}, q_i^{(u)} \geq 0 \quad i = 1, 2 \ldots, I \quad (5.17)$$

5.5 Numerical Examples and Test Problems

The following section illustrates the above-noted optimization model by solving a test example of a grocery category. The mixed-integer model features an objective function with linear constraints. The case examples will show that the computation time remains within reasonable time bounds.

Subsection 5.5.1 provides an overview of the test example, followed by the analyses of impact on total profit level in 5.5.2 and influence on solution structure in 5.5.3. Subsection 5.5.4 provides error bounds for parameter estimates and impact of hierarchical planning decisions with sensitivity analyses of the test example. In the final subsection 5.5.5, the model is tested with a large test case of a category with 200 items. The final test will evaluate the solution performance with a commercial solver and its applicability to other category sizes.

5.5.1 Description of the Test Case

5.5.1.1 Data Applied in Test Case

This section uses test problems with a wide range of parameter values for the computational experimental set-up. To investigate the effects of the problem, data from a single category are analyzed. The model requires retail outlet data as sales data, items facings, and the available shelf space. Item-specific data are item prices and costs, listing costs and item sizes. Ranges and constraints need to be determined for facings and basic supply levels. Real data from a single category of a retailer are used.

The experimental setting consists of 25 consumable products representing the entire product category. We use annual demand figures. The products are located within one shelf and regularly refilled, where $RF = 300$. The basic supply level is supposed to satisfy 50% of demand needs, i.e., only one additional refilling during

Table 5.2 Data for assortment, shelf space and replenishment test problem

Parameter	Description, values
α_i	Random distribution with average sales of 10,387 units and standard deviation of 13,811
c_i	Random distribution with $0.59 \le p_i \le 1.06$
k_i^{max}	max $[m_i \cdot 4; K]$ and rounded down to integer values
k_i^{min}	min $[m_i/4; 0]$ and rounded up to integer values
h	10%
LC_i	200 per item
r_i	Random distribution with $1.59 \le r_i \le 2.99$
RFC	2.50 per slot

the period is possible. Listing also needed to be included for required changes at the shelf layout. The space elasticity is assumed to be the same as it was in prior research (e.g., Hansen and Heinsbroek 1979; Yang 2001), i.e., $\beta_i = 0.2$ and consequently λ_j of 0.8. An exogenous substitution estimate is applied. The substitution shares are applied that the first alternative receives 50% of the substitution volume, and the second alternative 30%. The total substitution volume is limited to $\sum_{j=1}^{I} \mu_{ji} + n_j = 1$, with $n_j = 0.2$. Up to ten facing levels are allowed in the test. The data applied are summarized in Table 5.2.

5.5.1.2 Comparison of the CASRP Solution with Other Approaches

The objective value of the Capacitated Assortment, Shelf space and Replenishment Problem (CASRP) as presented in subsection 5.4 is compared with (1) a Base model (BM), and (2) a Capacitated Assortment and Shelf space Problem (CASP) as presented in Chap. 4:

- (1) The BM represents a lower bound of the objective value by applying the proportional-to-sales allocation rule, as it is usual in planogram software packages like "Apollo" or "Spaceman". Restocking-dependent costs are calculated a posteriori based on the specification of the decision variables. Additionally, substitution demand for delisted items has been redistributed to other items according to the provided rules in the test description.
- (2) The CASP solution is a shortened form of the objective function which does not take into account TCUS and TCOI. The replenishment and inventory costs are calculated a posteriori.

5.5.2 Profit Impact of Integrated Assortment, Shelf Space and Inventory Planning

The test cases with 25 items demonstrate the profit impact compared to current industry practice, and the benefit of an integrated restocking and shelf space model.

5.5 Numerical Examples and Test Problems

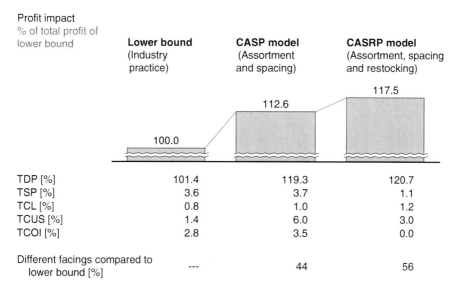

Fig. 5.3 Profit impact of integrated assortment, shelf space and inventory planning

The total profit increases by 17.5% compared to industry practice, which could be quite substantial in low-margin industry retailing. The CASP model optimizes only for listing and spacing, and reflects restocking requirements with basic supply level constraints, i.e., supply costs are not part of the objective function. Here, the test case shows a posteriori calculated raising over- and undersupply costs of the items. Disregarding refilling costs in space allocation results in partially frequent restocking requirements for some items and neglecting inventory holding costs results in high overstocks for other items. Therefore, both TCUS and TCOI are higher for the CASP model and the total profit is 4.4% lower (Fig. 5.3).

5.5.3 Impact on Solution Structure

The numerical example also show that the solution structure (i.e., facing levels) changes significantly. 56% of the items receive different facing levels in the optimized CASRP model compared to the lower bound.

The integration of restocking and inventory holding cost considerations impacts not only the total profit, but significantly changes the structure of the result as well, i.e., characteristics of the decision variables. Figure 5.4 summarizes the differences of the decision variables for each item. In comparison to the lower bound, applying a pure assortment and space optimization (CASP model), 44% of the items get different facing values assigned compared to industry practice. That means also, that the proportional rule assigns non-optimal facing values in around 50% of

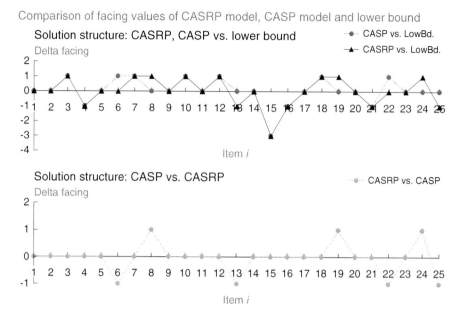

Fig. 5.4 Impact of inventory-related costs on solution structure

the cases. Integrating additionally the restocking implications as in the CASRP results in 56% of the items requiring different facing values compared to the lower bound.

Comparing the CASP model with the CASRP model shows the following results: Around 28% (7 of 25) of the items get different facing values. First of all, the integration of inventory holding costs forces to avoid high overstocked inventory. Figure 5.5 shows the four items where the facings levels decrease from five to four. There have been at $k = 5$ higher overstocked inventory levels.

Reducing the space of this four items creates space for other items. The CASRP model lists three more slow-mover items with a better supply-demand ratio. Figure 5.6 shows the demand- and supply-curves of the additionally listed items. Here the number of facings increase from zero (i.e., delisted) to one facing. In all three cases the facing-dependent supply and demand points are quite close and a limited volume need to be refilled.

5.5.4 Sensitivity Analyses of CASRP

A multitude of parameters need to be estimated. Errors and deviations cannot easily be excluded, so the impact of these parameters on the results of the model is evaluated using sensitivity analyses. In this subsection, sensitivity analyses are

5.5 Numerical Examples and Test Problems

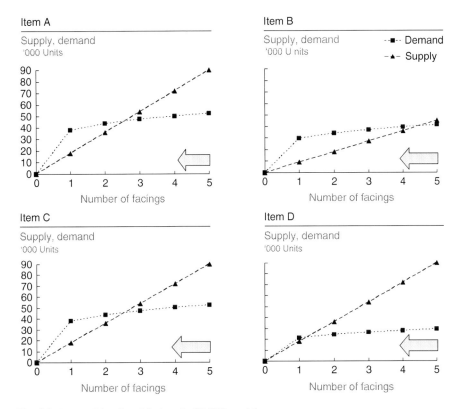

Fig. 5.5 Items with reduced facings in CASRP model

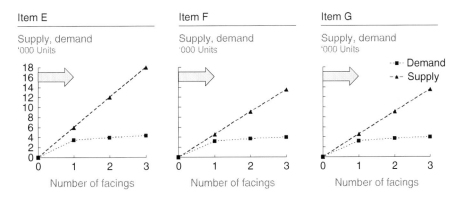

Fig. 5.6 Additionally listed items in CASRP model

computed using 10 levels of parameter changes: five with lower parameter values (-25% to -5%) than the default, and five with higher values ($+5$ to $+25\%$). The data used is identical to that of the previous section. The results are compared with relative to base profit, with no changes in the parameter values. The impact of hierarchical planning decisions, inventory-related costs and consumer behavior is investigated.

5.5.4.1 Hierarchical Planning Aspects

The determination of total shelf sizes and basic supply levels are managerial decisions based on hierarchical planning aspects. The variation in total shelf size significantly impacts objective value and restocking costs, as illustrated in Fig. 5.7. Reducing the shelf space by 25% results in a 31% lower overall profit. This is based on a lower profit due to lower sales, and especially the high costs of undersupply. The undersupply costs increase exponentially with high restocking requirements of low shelf sizes. Changes in the basic supply level have only a moderate effect on total profit by increasing the levels. Increasing the BSL level results in situations where a certain demand can no longer be satisfied, and so forces a change in shelf layout. Lowering the levels has a very limited profit effect, as a similar number of refilling processes are still required as in the test example.

Going forward, such sensitivity analyses can be used as feedback on the overarching planning question of total shelf space, and input to the subordination planning of basic shelf supply level for instore planning. A comprehensive planning framework and hierarchy facilitates the accuracy of decision making in retail sales planning as in Chap. 2.

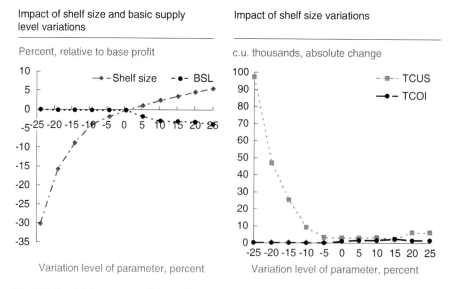

Fig. 5.7 Sensitivity analyses of the CASRP for managerial planning aspects

5.5.4.2 Inventory-Related Cost Parameters

The restocking costs need to be estimated using empirical studies, such as a time and motion observation of individual refilling processes. Interest rates signal opportunity costs. Although the replenishment costs have a significant impact on solution structure and overall profit, the variation in inventory-related cost parameters has a very limited impact on objective value. Higher or lower refilling costs and interest rates impact the total profit relative to the base profit by only $+0.4\%$ to -0.2%. Hence, the low profit sensitivity would also allow the use of cruder cost estimates.

However, variation in refilling costs impacts the listing decision. At first glance, the results on the right-hand chart of Fig. 5.8 for absolute profit impact by variation in the refilling costs are counterintuitive. Lower refilling costs result in smaller assortment sizes. Lowering the refilling costs by more than 5% results in delisting one item, and higher facings of another item. This has the following impact: (1) higher TSP from higher substitution volume; (2) lower TCL, as fewer products are listed; (3) no TCOI, as all items are undersupplied; (4) higher overall TCUS to serve substitution and direct demand.

The effect on TCUS appears particularly counterintuitive, as the per unit refilling costs decrease. However, what happens is simply that a high-demand item receives an additional facing. With lower per unit refilling costs, it even becomes more profitable to refilling this item via additional individual restocking than keep the other lower volume item, which did not be refilled. This example again shows that replenishment costs need to be integrated in assortment and shelf space management.

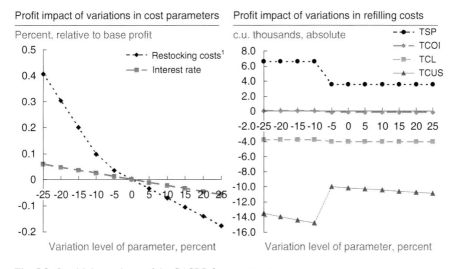

Fig. 5.8 Sensitivity analyses of the CASRP for cost parameters

5.5.4.3 Consumer Behavior

The parameters for consumer behavior (β_i, μ_{ji}) are estimated using consumer experiments and may thus by their very nature deviate from final consumer behavior in the stores and influence the results. The profit sensitivities for varying consumer behavior in space elasticity and substitution lie between -2.9% and $+2.3\%$ for a potential error range of -25% and 25%, as illustrated in Fig. 5.9.

The sensitivity analyses of the space elasticity effects show that disregarding restocking costs overestimates the profit potential of higher space effects and vice versa. Including restocking requirements in the objective functions lowers the impact of space-elastic demand, i.e., restocking costs work as buffer for over- or underestimates of space elasticity. Without restocking costs, as in the CASP model, the variation in space elasticity impacts profit by -2.9% to $+2.3\%$. With restocking, as in the CASRP model, the profit impact is -1.9% to -1.8%.

Variations in substitution intensity have only a very limited effect on the total objective value ($-0.1-+0.1\%$), but they have an impact on the solution structure and jump in intervals whenever additional products are (de-)listed. The effects are twofold:

1. The increasing substitution intensity shows that the overstocked volume decreases (see TCOI curve) and the TSP increases.
2. The decreasing substitution intensity of more than -10% shows that one product gets delisted as the substitution demand required to keep this item in the assortment becomes too low.

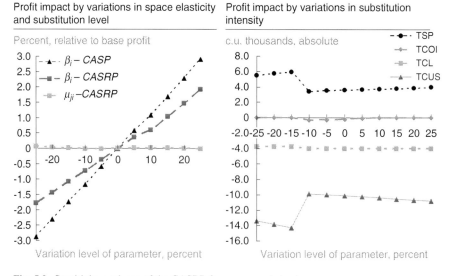

Fig. 5.9 Sensitivity analyses of the CASRP for consumer behavior

Overall, the sensitivity analyses provide valuable insights for profit implications and shelf space management decisions. Even though the effect of managerial decisions, cost estimates and expected consumer behavior might be low for the total profit, it has a significant impact on assortment and shelf space assignment decisions. Hence, these analyses also reveal the benefit of integrating replenishment considerations into the objective function for category planning to avoid disadvantageous decisions.

5.5.5 Applicability of CASRP for Large-Scale Categories

One should note that the scale of the optimization problem depends on the values of two variables: number of product items (I) in the category, and number of facing levels (K). For further numerical analysis therefore "large-scale" problem cases are chosen that include very high values for the variables, to ensure that the model is capable to solving also harder problem sizes a retailer might potentially face. For instance, a large supermarket retailer carries from approximately 35,000–50,000 items within 600 categories, i.e., the average number of distinct items in a category is around 60–80 (EHI Retail Institute 2010). The direct information from German retailers suggests that there are no categories exceeding 200 items (i.e., $I = 200$). The approach to generating random problem instances is similar to that adopted by Yang (2001), Lim et al. (2004) and Murray et al. (2010). Data are applied as in Chap. 4 to test a hard knapsack problem (Pisinger 2005). A profit-to-weight correlation with $R^2 = 0.8055$, random distribution of observed demand with average of 10,348 units and standard deviation of 10,390, and a total shelf size is applied that the previous sales fits into the shelf. The variation in facing levels is also restricted to currently existing shelf layout. Retailers prefer to change shelf layout incrementally so as not to confuse consumers. Therefore, facing variations are allowed that still fulfill lower and upper bounds as described in Table 5.2. This constraint is only relaxed in the test case with $K = 20$. μ_{ji} is set to 0.5 for the first alternative and to 0.3 for the second alternative. The share of lost sales is $\eta_j = 0.2$. A BSL of 0.5 is applied, so that only maximum one refilling during

Table 5.3 Evaluation of computation performance of the CASRP-model

I	K	ΔP to BM	Calc. time (s)
50	5	8.2%	<1
50	10	8.2%	<1
100	5	9.2%	<1
100	10	9.6%	<1
200	5	8.6%	6
200	10	9.4%	13
200	20	11.2%	9

the period is allowed. The total profit of the models is compared to the base model as in commercial shelf space software.

The computation times show that the extended model is capable of handling even large number of items and all relevant category problem sizes with very fast computation time. The CPLEX solver already found also for larger problem sizes the optimal solutions also after a few seconds (Table 5.3).

5.6 Conclusions and Future Areas for Research

Traditional approaches in shelf space management abstract from actual instore replenishment processes, where merchandizers execute group replenishment and sales staff deal with individual product replenishment.

The model differentiates the replenishment processes and integrates the inventory-related costs. It therefore extends shelf space management models by replenishment costs, and clarifies restocking requirements. It has been resolved as a mixed-integer problem in CPLEX, and allows the computation of optimal results for category-specific problem sizes. The numerical examples show the benefits of an integrated model over currently available software and shelf space management models.

The numerical examples prove that demand-related facing decisions require strong alignment with instore operations. The results show that a proportional space-to-sales rule, as applied in standard planogram software tools, tends to delist too many items and omits the impact on restocking costs. Neglecting restocking costs, as in traditional shelf space management models, also results in high replenishment frequencies. Sensitivity analyses indicate the need for hierarchical and integrated planning, especially for determining the overall shelf size of a category. Although the cost estimates and expected consumer behavior have a limited effect on total profit, variations in their level result in different shelf space decisions.

Areas of potential further research are the investigation of joint optimization of space assignment, instore replenishment cycles, and order cycles for backroom replenishment. Here, a further constraint to overall replenishment capacity related to basic filling and related to product-specific refilling processes could also be applied, if retailers have constrained overall workforce capacity. Furthermore, backroom capacity and inventory costs could be integrated, as only showroom effects and costs were reflected. Competitive scenarios and marketing effects such as pricing additionally influence demand, and could be part of an integrated analysis, as well as an extension of stochastic demand models.

After extending the CASP-model with supply effects, we will analyze additional demand effects by integrating pricing into shelf space management in the following chapter.

Chapter 6
Assortment, Shelf Space and Price Planning

6.1 Introduction and Motivation

Assortment, shelf space and price planning are both extremely important and challenging for retailers. Retailers need to satisfy consumer demand with shelf supply by balancing variety (number of products) and service levels (number of items of a product), and offer competitive prices with respect to limited and valuable operational resources such as shelf space and merchandizers. For example, offering broader assortments may limit appropriate service levels and vice versa. Lower item prices diminish the profit contribution per item sold, but increase demand and impact overall profitability, depending on price elasticity. Additionally, higher demand requires more frequent restocking or a higher number of facings. Retailers therefore benefit from comprehensive shelf space management. Several empirical studies show that well-executed shelf space management results in profit growth because of an efficient response to consumer behavior and effective category management (McIntyre and Miller 1999); (Hennessy 2001; Basuroy et al. 2001; Dhar et al. 2001; Shugan and Desiraju 2001; ECR Europe 2003a; Levy et al. 2004; Grocery Manufacturers Association et al. 2005; Desrochers and Nelson 2006; Campillo-Lundbeck 2009; Gutgeld et al. 2009; Fisher and Raman 2010). As a result, category management needs to carefully plan assortment size, shelf space assignment of products and price of each product.

The goal of this chapter is to addresses an integrated shelf space allocation, assortment and pricing problem by capturing the decision trade-offs faced by retailers in optimizing their shelf space. This chapter structures the planning problem, provides a decision support model to maximize category profit, and tests the capability of this model for solving a category-specific problem. Specifically, this chapter develops a model that jointly optimizes retailers' decisions relating to assortment size, product prices and number of facings in a product category, and captures cross-product demand interactions through substitution.

The remainder of this chapter is organized as follows: The second section establishes the context of the decision problem and reviews empirical insights, while Sect. 6.3 analyzes state-of-the art optimization models that are related to the planning problem being considered. Section 6.4 presents a novel model that jointly optimizes retailers' decisions relating to assortment size, product prices and number of facings in a product category, and captures cross-product demand interactions through substitution. Section 6.5 provides numerical examples. The chapter concludes with Sect. 6.6, which also develops further areas for research.

6.2 Problem Definition

The traditional retailer shelf management tool is a planogram. Planogram software products can provide a realistic view of the shelves, and are capable of allocating shelf space according to simple heuristics such as allocating space proportionally to sales or profit. A notable problem of all these systems is that they disregard important demand effects such as space elasticity and product substitution. However, the major drawback of shelf space software is that it abstracts from the category manager's same-time decision problem to decide not only about space allocation, but also about which products to list and how to price them. The relevant planning problems, which have to be considered simultaneously, can be defined as follows:

- *Assortment planning*: Listing decisions based on consumer choice behavior and substitution effects.
- *Shelf space planning*: Facing and replenishment decisions based on space- and cross-space elasticity effects, limited shelf space and operational restocking constraints.
- *Price planning*: Pricing decisions based on price- and cross-price elasticity effects.

All these planning problems could be significantly supported by modeling and optimization methodologies that allow an integrated perspective on all relevant decision variables. Our purpose is therefore to develop a model for determining the assortment as well as allocating shelf space and prices for individual products within a fast-moving retail category. The optimization problem includes an objective function that maximizes category profitability, using merchandising variables as the decision variables, and incorporating various constraints. Consider a retailer who needs to select items from a set of products, $N = \{1, 2, \ldots, i, \ldots I\}$, within a category, define the number of facings, $\bar{k} = (k_1, k_2, \ldots, k_i, \ldots, k_I)$, and assign the price levels $\bar{r} = (r_1, r_2, \ldots, r_i, \ldots, r_I)$ of each item i. Listed items are denoted by the set N^+ and delisted items by N^-. Thus, $N^+, N^- \in N$, $N^+ \cup N^- = N$ and $N^+ \cap N^- = \emptyset$. The basic constraints are: store shelf capacity, product availability, and lower and upper bounds on facings and prices. Consistent with

6.2 Problem Definition

prior research, an efficient replenishment system is applied to avoid temporary stockouts.

To reflect various enterprise goals, the model receives input from the overarching strategic decisions on store layout and master category planning. The latter forecasts mid-term category demand and defines strategies of categories such as overall category space S, item price corridors (r_i^{min}, r_i^{max}, e.g., pricing all pack sizes of one product group similarly), and minimum and maximum facings (k_i^{min}, k_i^{max}). Requirements for operational instore shelf replenishment are anticipated as basic supply level (BSL) constraints. In contrast to the permanent assortment, the promotional assortment focuses on temporary items and prices. This is not however the focus of this investigation. The analyses focus on a stable market environment where retailers optimize profit in a given competitive market situation. The disregard of direct competitive reaction reflects mid-term optimization for a period where e.g., prices and planograms of other retailers are already determined. The model accounts for gains from/losses to competition to the extent of a specific gain/loss sales factor.

The shelf space allocation and pricing problem are quite different depending on whether one takes the perspective of a retailer or a consumer goods producer (see Fig. 6.1). Producers want to maximize only the sales and profits of their own products. They always want more and better shelf space for their products, and only care about store switching in a limited extend. In contrast, retailers want to maximize category sales and profits, regardless of producer shares. The following model is formulated from the perspective of category management by retailers, but could be adjusted for a producer if the relevant product and cross-product relations were taken into account.

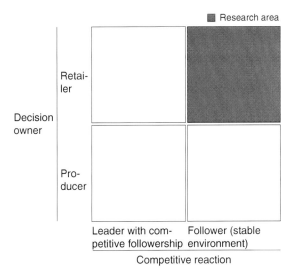

Fig. 6.1 Decision ownership and interaction with competition

Excursus: Pricing Perspectives

Prices can usually be set from three perspectives: cost-plus pricing, value-based pricing and competitive pricing. The first is the most basic method, setting prices based on the input costs and a target profit margin. This is more an accounting view. The isolated view does not integrate consumer reactions, which takes place in the second method. In value-based-pricing consumer demand is the core of price definition. Consumer price sensitivity, i.e., the price-demand curve, is used to maximize consumer and producer values. Consumer demand reaction is anticipated with alternative prices. Competitive pricing has its origin in industrial organization, where a competitive reaction function is used to anticipate competitors' price changes in an oligopolistic market for example. Consequently, an integrated model needs to fulfill all these three perspectives by matching input costs and profit targets with consumer and competitor reactions.

6.2.1 Properties of Demand Function

A further objective of this research is to investigate and integrate some of the most relevant instore consumer choice effects into the optimization model. We therefore introduce empirical research and use this to formulate the associated demand effects. The majority of consumers make their final purchase decisions instore, have a low level of involvement with their instore decisions, and make choices very quickly after a minimal search (Drèze et al. 1994; Xin et al. 2009; Chandon et al. 2009). The demand model expresses these facts formally.

6.2.1.1 Space-Dependent Demand

Shelf space allocation influences consumer attention and demand. Drèze et al. (1994), Campo and Gijsbrechts (2005) and Inman et al. (2009) verify that fast-moving consumer goods have significant space elasticity. Chandon et al. (2009) reveal that facing variation is the most significant instore factor.

Common denominators of shelf space models are item demand rates as a function of the space allocated to each item. The basic demand is denoted by α_i, β_i describes the space elasticity, and b_i is the breadth of item i. Common retailer practice is to move items forward to the front row. The space-dependent demand $\tilde{d}_{ik}$ of item i is a deterministic function of its displayed front-row inventory level at the facing levels k (Hansen and Heinsbroek 1979).

$$\tilde{d}_{ik} = \alpha_i \cdot (k \cdot b_i)^{\beta_i} \qquad i = 1, \ldots, I; k = 1, \ldots, K \qquad (6.1)$$

6.2.1.2 Price-Dependent Demand

Secondly, Hoch et al. (1994) demonstrate the high impact of retail pricing practices on profits. Their experiment results in up to 32% category profit impact for price changes. Tellis (1988) and Gaur and Fisher (2005) also describe the effect of pricing on consumer demand. Such evidence is consistent with retail shoppers' propensity to compare prices (Bucklin et al. 1998; Urbany et al. 2000; Bimolt et al. 2005; Chandon et al. 2009).

The price-dependent demand function describes the relationship between price and aggregated consumer demand: The lower the price, the higher the demand, and vice versa. Price elasticity ϵ_i for product i, where $i = 1, 2, \ldots, I$, represents the demand change relative to the price change. Retail prices are rounded to a set of discrete price points. It is therefore not over-restrictive to limit this case to a set of discrete price points, $r_i \in R_i$. Further, to allow a high degree of price flexibility, an elasticity by price point ϵ_{il} is proposed, where $i = 1, 2, \ldots, I; l = 1, 2, \ldots, L$, as this elasticity may vary by price point, e.g., when a threshold is reached. The price levels are denoted by index l, where $l = 1, 2, \ldots, L$. The price index n denotes the price level observed, i.e., the base price, and α_i defines the associated demand for item i at base price level n. The price-dependent demand $\hat{d}_{il}$ of item i at price level l is formulated as follows:

$$\hat{d}_{il} = \alpha_i \cdot \left(1 + \epsilon_{il} \cdot \frac{r_{il} - r_{in}}{r_{in}}\right) \qquad i = 1, \ldots, I; l = 1, \ldots, L \qquad (6.2)$$

Combining space- and price-dependent demand effects leads to the following demand rate of item i, d_{ikl}, which depends on the number of facings k and the chosen price level l:

$$d_{ikl} = \alpha_i \cdot (k \cdot b_i)^{\beta_i} \cdot \left(1 + \epsilon_{il} \cdot \frac{r_{il} - r_{in}}{r_{in}}\right) \qquad i = 1, \ldots, I; k = 1, \ldots, K; l = 1, \ldots, L$$
$$(6.3)$$

6.2.1.3 Assortment-Dependent Demand

Demand decisions relating to consumer goods cannot be taken without considering the overall impact of other items, i.e., with respect to listing, spacing and pricing. These decisions – assuming space is limited – imply that potentially not all products can be added to the assortment, for example, or that it may be more profitable to list other products to force consumers to switch to more profitable substitutes. There is also empirical evidence that variety levels have become so excessive that reducing variety significantly increases sales (Sloot and Verhoef 2008). Also, Drèze et al. (1994), Iyengar and Lepper (2000) and Dhar et al. (2001) report a positive impact from reducing assortment size and delisting items on total demand. Boatwright and Nunes (2001) found that significant item reductions (up to 54%) resulted in an average sales increase of 11% across 42 categories examined, and sales growth in more than two-thirds of these categories. Gruen et al. (2002) report that 45% of consumers substitute, i.e., buy one of the available items from the same category. ECR Europe (2003b) concludes that 69% of the volume is substituted, while the research of Xin et al. (2009) states that the relevant figure is up to 75%, and the work of van Woensel et al. (2007) cites a figure of 84%.

Thus, we follow the general idea of classical assortment approaches in using exogenous substitution estimates. Every consumer chooses their favorite item j from set N. Substitution demand is determined by the substitution intensity and latent demand for delisted items $\tilde{d}_{j0}$ ($j \in N^-$). If their favorite product j is not available for some reason, probability μ_{ji} approximates that a consumer will choose a second favorite i, $i \in N^+$. The fraction of consumers who are willing to compromise their initial choice for product j is expressed in the probability of $(1 - \eta_j)$, with $\eta_j = 1 - \sum_{\substack{i=1 \\ i \neq j}}^{I} \mu_{ji}, j = 1, 2, \ldots I$. Note that η_j does not depend on the stocking level due to the assumption of an efficient replenishment system.

We assume as in the CASP-model the latent demand of delisted items with $\tilde{d}_{j0} = \lambda_j \cdot \tilde{d}_{j1}$ $j = 1, \ldots, I$ and one round of substitution.

6.2.1.4 Cross-Space- and Cross-Price-Dependent Demand

Cross-product effects also need to be an integrated part of the model. Consumers substitute, i.e., buy alternatives if the changes to price and space increase the attractiveness of the alternative. Cross-space and cross-price elasticity quantify the effects of facings and prices of alternative items j on the demand for an item i. In the past, every consumer has chosen their favorite item j, $j \in N^+$, from all listed items. If their favorite item now has varying shelf facings or shelf prices, the consumer will switch to a second favorite with a probability of μ_{ji}. Here we also use an exogenous demand substitution matrix. These effects are cannibalization and gains from other items dependent on their facings and prices. The substitution may be different for demand substitutions away from items versus substitutions to items. That is why the substitution matrix does not need to be symmetrical ($\mu_{ji} \neq \mu_{ij}$).

The following Fig. 6.2 illustrates the demand shifts for price-related demand on the left hand side. The demand for an individual item is precalculated for discrete prices, whereas shifts between items are reflected in the extent to which substitutions are made. The right chart illustrates the demand shifts for changes in the facing levels accordingly.

6.2.1.5 Other Demand Effects

The model focuses on optimizing the permanent assortment. Promotional and other marketing activities are not part of the decision model. Effects other than space allocation and regular shelf prices are assumed to be constant. The demand

6.2 Problem Definition

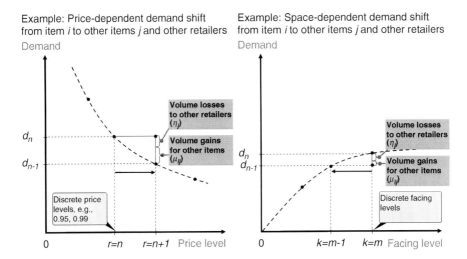

Fig. 6.2 Example: Cross-product demand shifts

function relates to products with clearly defined qualities. An improvement of the product quality would lead to a shift of the curve. The demand function assumes a specific constellation of the environment, especially regarding consumer behavior, competitive reaction and stability of the factors influencing behavior. This premise allows the abstraction of uncertainty from pricing decisions, and the use of deterministic models. Dynamic pricing systems such as varying prices within the period, life cycle pricing and temporary price reductions will not be integrated in the basic model either. This also excludes mixed calculation with loss leaders. The demand function will refer to a clearly defined period. It will thus be a static model. This implies that time-related substitution, i.e., carry-over effects of decisions in a period, do not impact demand in later periods. Other price differentiation systems, such as personal, quantitative, or price bundling, will not be applied as they are rather unusual for everyday pricing and regular store assortments.

Complementary effects can easily be integrated by adjusting the substitution matrix to negative values.

6.2.1.6 Total Demand Function

Our approach makes it possible to calculate the total item demand d^*_{ikl} by combining all demand effects related to the total demand function for a listed item i, $i \in N^+$. It consists of the facing- and price-level-dependent demand d_{ikl}, the substitution demand from delisted items N^-, $\sum_{j \in N^-} \tilde{d}_{j0} \cdot \mu_{ji}$, the demand gain d_i^+ and demand cannibalization d_i^- via changes in the facing and/or price level of other listed items j, $j \in N^+$, $j \neq i$.

$$d^*_{ikl} = d_{ikl} + \sum_{j \in N^-} \tilde{d}_{j0} \cdot \mu_{ji} + d^+_i \left(r_j, k_j, \forall j \in N^+\right) + d^-_i \left(r_j, k_j, \forall j \in N^+\right)$$
(6.4)

6.2.2 Instore Inventory Management and Shelf Replenishment

Instore handling costs amount from 38% to 48% of operational retail logistics costs (Broekmeulen et al. 2006; Kuhn and Sternbeck 2011), often because shelf space allocation is not aligned with the replenishment regime (Ketzenberg et al. 2002; Curseu et al. 2009; van Zelst et al. 2009). The analyses in Chap. 5 shows that up to 70% of the items are temporarily underfaced, i.e., consumer demand is higher than shelf supply, thus requiring frequent restocking. Traditional shelf space decision models disregard this effect by assuming that each demand can be fulfilled (Hansen and Heinsbroek 1979; Corstjens and Doyle 1981; Urban 1998; Hariga et al. 2007; Abbott and Palekar 2008; Hansen et al. 2010; Murray et al. 2010). We prefer to follow actual retail practice by positing a differentiated replenishment system: (1) all items are replenished together at regular intervals using scheduled basic filling with a fixed quantity, e.g., by merchandizers before the store opens; (2) individual replenishment of items by sales employees during the day as soon demand outstrips shelf supply. We therefore integrate operational replenishment requirements into the strategic model and introduce a "basic supply level" (BSL), which is a percentage of demand that is covered by the basic filling process. This level needs to be achieved by the regular daily basic shelf-filling process, whereas sales staff can fulfill the residuum during the day. A BSL of 100% means that total demand is fully satisfied by the scheduled basic stocking. A lower BSL requires limited individual restocking. It is assumed that retailers employ an effective logistics system, ensuring that demand required in the showroom can be replaced immediately with stock available from the backroom. This restocking regime avoids temporary out-of-stock situations, as the entire demand can be fulfilled by items available on the shelves, i.e., by basic filling and limited refilling.

6.3 Literature Review of Shelf Space and Price Planning

The following section summarizes the relevant literature on decision support models for shelf space planning, and puts the contribution of this study in this context.

Among the existing models in the shelf space management literature, one of the earliest studies is by Hansen and Heinsbroek (1979). In their model, demand for each product is a function of space elasticity. Their model seeks to maximize retailers' profits subject to a limit on total shelf space, upper and lower bounds on individual product quantity, and integer values for facings. Further,

6.3 Literature Review of Shelf Space and Price Planning

Zufryden (1986) introduces a dynamic programming formulation for a problem with space elasticity and demand-related marketing variables, including price. Although they have not considered cross-effects, this is the first shelf space model to include price-dependent demand effects. Yang and Chen (1999) assume a linear profit within a constrained number of facings within lower and upper bounds. Yang (2001) proposes a knapsack heuristic for the model. He found an optimal solution only for simplified versions. Also, the Yang model does not reflect cross-product effects.

Corstjens and Doyle (1981) developed a space allocation model with demand as a function of direct space and cross-space elasticity. A major criticism is that the model requires a pre-defined assortment, as delisting one product would result in zero demand for the entire category. In addition, the models do not assume integer facing values. Furthermore, they allow only a limited scope of items, and also assume price as an exogenous parameter. Borin et al. (1994) differentiate "unmodified," "modified demand," "acquired" and "stockout" demand. Unmodified demand is basic demand reflecting consumer preference for an item, whereas modified demand is solely a function of space allocation, such that price effects are ignored. Acquired and stockout demand reflect demand for out-of-assortment and out-of-stock substitution. Besides the exclusion of space elasticity, they also neglect operational constraints. Irion et al. (2004) further extend Corstjens' model to a product level instead of category level. Using a linearization framework, they transform the model into a mixed-integer problem with linear constraints. Their approach provides near-optimal solutions with a posteriori error bound. They simplify by using the identical demand for listed and delisted items ($\tilde{d}_{i0} = \tilde{d}_{i1}, i \in N$), and thus do not account for assortment decisions with latent demand.

Most literature on retail stores focuses primarily on the demand side, and less on the cost side. However, retailers with limited space face a trade-off of putting fewer items out for sale against keeping inventory of other products. Urban (1998) develop an inventory-theoretic approach to shelf space allocation and assortment decisions. The author combines order quantity decision, shelf space assignment and assortment decision into one model. Replenishment is assumed instantaneously and individually for each product. It therefore does not reflect actual retail policies.

Martín-Herrán et al. (2006) examine shelf space allocation and wholesale prices with a simulation in a static game with a Stackelberg equilibrium, with two manufacturers of two competing brands at one retailer. The manufacturers optimize their prices by taking into account shelf space allocation and price markup decisions of the common retailer. However, they exclude cross-space effects and adopt a cost-based pricing approach as they apply the same markup for the two brands. They derive optimal wholesale prices in the limited study, which features just two products. Murray et al. (2010) develop a model that jointly optimizes prices, facings, display orientation and shelf selection in a category. They apply a branch-and-bound-based MINLP algorithm that is able to solve problems in a fast and practical manner. The authors integrate space, price and cross-price effects, but disregard operational constraints and cross-space effects. The assortment also needs to be specified before.

While the above-noted models have made significant progress in addressing shelf space decision problems, they fall short in capturing the key aspects of the category manager's decision problem in their assumptions on:

- Pricing and/or assortment as given parameters.
- Immediate shelf replenishment and disregard of operational constraints.

A drawback of these studies is their (non-) applicability to assortment decisions. Generally, the models handle assortment decisions and latent demand for non-listed items (i.e., facing set to zero) by assuming no demand substitution, except for the models by Borin et al. (1994) and Urban (1998), and the CASP-model. A further limitation of existing models is their failure to consider a retailer's pricing and shelf allocation decisions jointly. Exceptions are the studies of Zufryden (1986), Martín-Herrán et al. (2006) and Murray et al. (2010), who reflected price elasticity, but without assortment effects. Additionally, all models allow immediate shelf replenishment to avoid stockouts. This inventory assumption can result in situations where retailers are forced to restock frequently and face substantial operational costs of items where demand is high and shelf space low. All is all, this is too divorced from the reality of the decision problem in retail practice, where retail category managers define shelf space, assortment and prices interdependently.

Excursus: Review of Other Pricing and Demand Planning Streams

A further review of streams related to the topic of integrated shelf space and price management indicates either a qualitative character, focus on temporary mark-down pricing, or assumes unlimited shelf capacity.

For example, ECR initiatives aim to match concepts of (supply-side) supply chain management with (demand-side) category management using qualitative approaches to strengthen the integrated view (Efficient Consumer Response 1993; ECR Europe 2003a; Fisher and Raman 2010).

Revenue management models study integrated price and inventory management. These models deal with dynamic and mark-down pricing for promoted or perishable items, i.e., time-varying prices (Gallego and van Ryzin 1994; Federgruen and Heching 1999; Kambil and Agrawal 2001; Bitran and Caldentey 2003; Quante et al. 2009). Furthermore, price promotion studies evaluate and optimize temporary price promotions (Gupta 1988; Abraham and Lodish 1993; Blattberg and Neslin 1993; van Heerde et al. 2004). They focus on highly varying prices between micro-periods (van Heerde et al. 2004), and not on mid-term shelf prices for the permanent assortment, which is the core of this contribution. Finally, price models assume unlimited shelf supply and omit shelf capacity and restocking costs, see e.g., Binkley and Connor (1998).

6.4 Formulation of the Capacitated Assortment, Shelf Space and Price Problem (CASPP)

In this section we develop an MIP model that addresses the integrated Capacitated Assortment, Shelf space and Price planning Problem (CASPP). It is a model for determining the assortment as well as allocating shelf space and prices for individual products within a fast-moving retail category. The optimization problem can be summarized with an objective function that maximizes category profitability, using merchandising variables as the decision variables, and incorporating various constraints. Consider a retailer who needs to select items from a set of $i = 1, 2, \ldots I$ products within a category, and where the demand d_i depends on number of facings k_i and price r_i, and the substitution demand from item j to i, where the demand for j depends on k_j and r_j. The item profit p_i depends on the selected price. The category manager needs to decide about the facings and price of each item to maximize the category profit P.

The composite demand function (6.4) can be transformed to a linear function by using the integer constraint for facing and discrete prices and precalculating the demand for each item. The non-linear demand function can then be transferred into a specialized knapsack problem where the binary decision variable y_{ikl} selects the most profitable item-facing-price combination for each item in the possible assortment. Hence, it can be degenerated into a bounded 0/1 multi-choice knapsack problem by precalculating all integer demand values, as there is a set of I items and each item i is associated with size b_i, price-level-dependent profit p_{il}, and total shelf capacity S. Additionally, the binary variable z_i denotes whether an item is listed or delisted.

6.4.1 Objective Function

The retailer's objective is to maximize product category profit P. The total profit comprises *total direct profit* (TDP), *total substitution profit* (TSP), *total cross-product profit* (TCPP) and *total costs of listing* (TCL).

$$\text{Max! } P = \text{TDP} + \text{TSP} + \text{TCPP} - \text{TCL} \tag{6.5}$$

6.4.1.1 Total Direct Profit (TDP)

The profit from individual product demand covers the profit of items regardless of their relation to the remaining assortment. The demand (6.3) is used to precalculate the demand d_{ikl} for each item. Specifically, the model decides how many facings at which price levels of each item are to be placed into the knapsack to maximize TDP.

$$\text{TDP} = \sum_{i=1}^{I} \sum_{k=1}^{K} \sum_{l=1}^{L} y_{ikl} \cdot d_{ikl} \cdot p_{il} \tag{6.6}$$

The profit p_{il} includes shelf prices r_{il} and unit costs c_i, i.e., $p_{il} = r_{il} - c_i$.

6.4.1.2 Total Substitution Profit (TSP)

Equation (6.7) describes the total substitution profit of item i assuming a given substitution volume d_i^{N-} from the set of items delisted, $j \in N^-$, to an item listed, $i \in N^+$. The substitution demand can only be realized and increase the demand for i if item i is listed, i.e., $z_i = 1$ and $y_{ikl} \neq 0$. The substitution volume of item i is given by (6.8). The parameter $\tilde{d}_{j0}$ quantifies the latent demand for delisted items and can be approximated as formulated in Sect. 6.2.1. The rate μ_{ji} accounts for the substitution from product j to i.

$$\text{TSP} = \sum_{i=1}^{I} \sum_{k=1}^{K} \sum_{l=1}^{L} d_i^{N-} \cdot y_{ikl} \cdot p_{il} \tag{6.7}$$

with

$$d_i^{N-} = \sum_{\substack{j=1 \\ j \neq i}}^{I} \tilde{d}_{j0} \cdot (1 - z_j) \cdot \mu_{ji} \quad i = 1, \ldots, I \tag{6.8}$$

6.4.1.3 Total Cross-Product Profit (TCPP)

The total cross-product profit (TCPP) integrates demand gain d_i^+ and demand cannibalization d_i^- via demand shifts between items j and i due to changes of the facings and prices of item j, i.e., cross-space and cross-price elasticity, with $j \in N^+, j \neq i$.

$$\text{TCPP} = \sum_{i=1}^{I} \sum_{k=1}^{K} \sum_{l=1}^{L} (d_i^+ + d_i^-) \cdot y_{ikl} \cdot p_{il} \tag{6.9}$$

The demand gain d_i^+ given in (6.10) describes that demand for item i increases if the alternative item j has either lower facings, higher prices or both. Consequently the demand for item j is lower and partially distributed to item i, defined by the substitution matrix μ_{ji}. Note that the indices m and n symbolize the number of facings observed and the price level observed, respectively.

6.4 Formulation of the Capacitated Assortment, Shelf Space and Price Problem (CASPP)

$$d_i^+ = \sum_{\substack{j=1 \\ j \neq i}}^{I} \left[\sum_{k=1}^{m-1} \sum_{l=1}^{L} (d_{jml} - d_{jkl}) \cdot y_{jkl} \cdot \mu_{ji} \right.$$

$$\left. + \sum_{k=1}^{K} \sum_{l=n+1}^{L} (d_{jkn} - d_{jkl}) \cdot y_{jkl} \cdot \mu_{ji} \right] \quad i = 1, \ldots, I \quad (6.10)$$

The left term in brackets describes the increase in demand for item i if facings of items j ($k < m$) are lower. The right term denotes the increase in demand for item i if prices of items j ($l > n$) are higher.

On the other hand, demand for item i falls if the alternative item j has either higher facings, lower prices, or both. Consequently the demand for item j is cannibalized by item i. The substitution matrix μ_{ij} describes the intensity of cannibalization. The effect is described by the demand cannibalization d_i^- of item i given in (6.11):

$$d_i^- = \sum_{\substack{j=1 \\ j \neq i}}^{I} \left[\sum_{k=m+1}^{K} \sum_{l=1}^{L} (d_{jml} - d_{jkl}) \cdot y_{jkl} \cdot \mu_{ij} \right.$$

$$\left. + \sum_{k=1}^{K} \sum_{l=1}^{n-1} (d_{jkn} - d_{jkl}) \cdot y_{jkl} \cdot \mu_{ij} \right] \quad i = 1, \ldots, I \quad (6.11)$$

The left term in brackets describes the demand cannibalization of item i by item j when facings of item j ($k > m$) are higher, the right term describes demand cannibalization of item i by item j when the price of item j ($l < n$) is lower. Note that $d_i^+ \geq 0$ and $d_i^- \leq 0$.

6.4.1.4 Listing Costs (TCL)

New product listings induce fixed costs of product advertising and changes in store layout. Listing costs LC_i occur if item i is listed.

$$\text{TCL} = \sum_{i=1}^{I} z_i \cdot LC_i \quad (6.12)$$

6.4.2 Constraints

The constraints are composed of hierarchical planning aspects, modeling requirements and demand effect boundaries. Constraints reflect the input from overarching decisions like instore layout or category planning. Consequently, the shelf sizes of categories, price and facing corridors or share/volume targets act as inputs to the optimization model. The following constraints are applied:

$$\sum_{i=1}^{I}\sum_{k=1}^{K}\sum_{l=1}^{L} y_{ikl} \cdot k \cdot b_i \leq S \tag{6.13}$$

$$k_i^{\min} \leq \sum_{k=1}^{K}\sum_{l=1}^{L} y_{ikl} \cdot k \leq k_i^{\max} \qquad i = 1, 2, \ldots, I \tag{6.14}$$

$$r_i^{\min} \leq \sum_{k=1}^{K}\sum_{l=1}^{L} y_{ikl} \cdot r_{il} \leq r_i^{\max} \qquad \forall i \in N^+ \tag{6.15}$$

$$\left[\sum_{k=1}^{K}\sum_{l=1}^{L} d_{ikl} \cdot y_{ikl} + d_i^{N^-} + d_i^+ + d_i^-\right] \cdot \text{BSL}$$
$$\leq \sum_{k=1}^{K}\sum_{l=1}^{L} y_{ikl} \cdot k \cdot g_i + (1 - z_i) \cdot d_i^{max} \quad i = 1, \ldots, I \tag{6.16}$$

$$z_i + \sum_{k=1}^{K}\sum_{l=1}^{L} y_{ikl} = 1 \qquad i = 1, 2, \ldots, I \tag{6.17}$$

$$y_{ikl} \in \{0; 1\} \qquad i = 1, 2 \ldots, I; \quad k = 1, 2, \ldots, K; \quad l = 1, 2, \ldots, L \tag{6.18}$$

$$z_i \in \{0; 1\} \qquad i = 1, 2 \ldots, I \tag{6.19}$$

Constraint (6.13) ensures that only the available space S can be distributed. (6.14) sets boundaries on the number of facings with a maximum (minimum) number of facings of item i by $k_i^{\max}$ ($k_i^{\min}$). The facing range depicts business restrictions like presentation of certain item types, enforces minimum listings for new products, or sets upper limits for items to only get a share of shelves. As noted in the literature (e.g., Yang 2001; Murray et al. 2010), the lower bound $k_i^{\min}$ captures the retailer's contractual obligations with producers that require that a minimum amount of facing area is assigned to each product item. (6.15) sets price corridors, with the maximum and minimum prices of a listed item as $r_i^{\max}$ and $r_i^{\min}$ based on relevant competitive market dynamics. It can also express that some products need to remain in a certain proportion to each other, e.g., $r_i^{\min} > r_j^{\max}, i \neq j$. The BSL in constraint (6.16) puts supply and demand in relation to one another. The left side covers direct demand and substituted demand of item i, while the right side denotes the supply capacity for basic shelf replenishment. The number of units supplied via basic refills per facing is described by g_i. Only listed items ($z_i = 1$) need to achieve the BSL and fulfill constraint (6.16). However, since values for substituted demand may be positive even though item i is not listed ($z_i = 0$), it is necessary to ensure the validity of the constraint for delisted items by adding a large number on the right side, e.g., the maximum demand for item i, d_i^{max}. (6.17) guarantees that only one facing level and one price can be selected for each item listed. (6.18) and (6.19) define the decision variables y_{ikl} and z_i as binary variables.

6.5 Numerical Examples and Test Problems

The following section illustrates the above optimization model by solving a test example from a grocery category. This section first provides an overview of the test example in 6.5.1, followed by the analyses of impact on total profit level in 6.5.2 and additional managerial insights in 6.5.3. Subsection 6.5.4 provides error bounds for parameter estimates and the impact of hierarchical planning decisions with sensitivity analyses of the test example. In a final subsection 6.5.5, the model is tested with a large test case from a category with 80 items. The final test will evaluate the solution performance with a commercial solver, and examine its applicability to other category sizes.

6.5.1 Description of the Test Case

6.5.1.1 Data Applied in Test Case

The numerical examples use test problems with a wide range of parameter values for the computational set-up for the experiment. To investigate the effects of the problem, observed data from a single retail category are analyzed. The model requires retail outlet data for item sales and their facings, and the available shelf space. Item-specific data are item prices and costs, listing costs and item sizes. Ranges and constraints need to be provided for facings, prices and basic supply levels.

The experimental setting consists of up to 30 consumer products representing an entire product category. Annual demand figures are used. The products are located within one shelf and regularly refilled. Direct shelf replenishment needs to cover 80% of the demand, i.e., BSL = 80%. Profit was modeled using item gross margin, as handling and replenishment costs are the same across the merchandize category for the retailer. Listing costs LC_i of 1.000 currency units also needed to be included for changes required to the shelf layout and other processes.

Price elasticity was provided by previous consumer research in this particular category of the retailer, with $-1.75 \leq \epsilon_{il} \leq -1.45, i = 1, 2, \ldots I; l = 1, 2, \ldots L$. The space elasticity is assumed to be the same as it was in prior research (e.g., Hansen and Heinsbroek 1979; Yang 2001), i.e., $\beta_i = 0.2$ and consequently $\lambda_i = 0.8$. For a sake of simplicity in the numerical analyses, we apply a symmetrical substitution matrix ($\mu_{ji} = \mu_{ij}, \forall j, i \in N$). An exogenous substitution estimate is applied, with $\mu_j^{(1)} = \sum_{i=1}^{N} \mu_{ji} = 0.8, \forall j \in N$, as the substitution volume for the first alternative of item j, and $\eta_j = 0.2, \forall j \in N$. We also apply bounds on the number of facings. The lower bound $k_i^{\min}$ (upper bound $k_i^{\max}$) is set at 25%

Table 6.1 Sales and profit data for assortment, shelf space and price test problem

Parameter	Description, values
d_{imn}	Annual demand observed, with average demand of 11,951 units and a standard deviation of 12,819 units
p_{in}	Uniformly distributed between $0.79 \leq p_{in} \leq 1.95$

(400%) of the facings observed m. The bounds on the price levels for listed items are: $r_i^{min} = 0.95 \cdot r_{in}$ and $r_i^{max} = 1.05 \cdot r_{in}$. The difference between each price levels is 2.5% and rounded to discrete values. In the tests, ten facing levels ($K = 10$) will be allowed and five pricing levels ($L = 5$). The observed sales data applied are summarized in Table 6.1.

6.5.1.2 Comparison of the CASPP Solution with Other Approaches

The objective value of the "Capacitated Assortment, Shelf space and Pricing Problem" (CASPP) with space-elastic, substitution and price-elastic effects as presented in this chapter is compared with (1) a base model (BM), (2) a "Capacitated Shelf space Problem" (CSP), and (3) a "Capacitated Assortment and Shelf space Problem" (CASP):

1. The BM represents a lower bound on the objective value by applying the proportional-to-sales allocation rule (without adjusting prices), as is usual in commercial software packages like "Apollo" or "Spaceman." Substitution is calculated a posteriori.
2. The CSP solution takes into account space and cross-space effects. The objective function is denoted by $P = \text{TDP} + \text{TCPP} - \text{TCL}$, substitutions for delistings are calculated a posteriori, and prices are exogenously specified (i.e., remain at base price r_{in}).
3. The CASP additionally integrates substitutions with $P = \text{TDP} + \text{TSP} + \text{TCPP} - \text{TCL}$. We specified prices exogenously (as above).

6.5.2 Profit Impact of Integrated Assortment, Shelf Space and Price Planning

The objective values of the models with varying problem sizes are compared with the base case solution of the BM. Table 6.2 shows increases in total profit for the CSP, CASP and CASPP models. The CASPP model increases profit by 6–23% over BM, and by 1–19% over the CASP solution, which could represent substantial potential in a low-margin industry like retailing. Also the CSP and CASP models significantly outperform standard industry practice by allocationg space proportionally.

6.5 Numerical Examples and Test Problems

Table 6.2 Profit impact of integrated assortment, shelf space and price planning

I	K	L	CSP	CASP	CASPP
5	10	5	3.2%	3.6%	22.7%
10	10	5	0.2%	5.2%	5.9%
15	10	5	4.2%	4.9%	10.6%
20	10	5	1.3%	5.2%	6.7%
25	10	5	0.8%	1.8%	5.8%
30	10	5	1.0%	6.6%	6.8%

6.5.3 Further Managerial Insights

In the following we analyze the example where $I = 20, K = 10, L = 5$ to generate managerial insights. The integration of price variation impacts not only the total profit, but significantly changes the structure of the result as well, i.e., characteristics of the decision variables. Figure 6.3 summarizes the differences in the decision variables for each item. Comparing the CASP model with the CASPP model leads to the following results: 50% of the products are assigned different prices, 56% receive different facings. This results in up to 60% of items getting different facings and/or prices via the integrated approach. This example shows that the possibility of changing prices significantly impacts solution structure and total profit.

Figure 6.4 shows that integrated optimization improves the utilization of shelf capacity and aligns supply and demand. Overall shelf space utilization (= space requirement of total demand/total shelf supply) increases significantly if varying prices are allowed in the integrated model. The shelf space utilized increases from 89% for the base case to 91% for the CASP case, and to 96% with the CASPP model. The CASPP model better leverages the available shelf capacity. A utilization of 100% means that all items were sold before basic restocking.

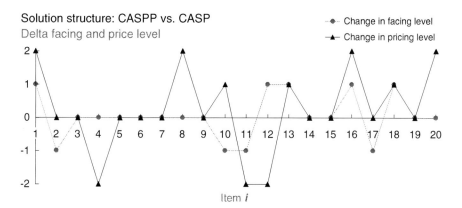

Fig. 6.3 Comparison of facing and pricing values in the CASPP and CASP models

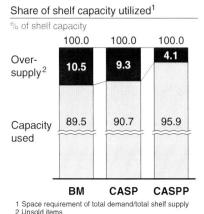

Fig. 6.4 Improvement of shelf utilization through the CASPP-model

6.5.4 Sensitivity Analyses

A multitude of parameters need to be estimated. Errors and deviations cannot be excluded easily, so an evaluation of the impact of these parameters on the results of the model is carried out using sensitivity analyses. In this subsection, sensitivity analyses are computed using four levels of parameter changes, where five have lower parameter values (-10–-50%) than the default, and five have higher values ($+10$–$+50\%$). The parameters for consumer behavior (β_i, ϵ_{il}, μ_{ji}) were estimated in consumer experiments, and may therefore by their nature deviate from final consumer behavior in stores, and influence the results. The determination of total shelf sizes and basic supply levels are managerial decisions based on hierarchical planning aspects.

The sensitivity analyses reveal moderate to medium impact on profit for all parameters, as illustrated in Fig. 6.5. The managerial planning decisions on category shelf space and desired BSL impact profit prospects. Assigning too little shelf space affects profit significantly. Increasing or lowering the basic supply level impacts profit, but with only limited effects. However, it shows that a comprehensive planning framework and hierarchy would facilitate the accuracy of decision making in retail sales planning. The profit sensitivity for varying consumer behavior in terms of elasticity and substitution lies between -2.2% and 2.3%, with a potential error range of -50% and 50%. This would allow the use of even rougher data sets. Any higher accuracy in estimating consumer behavior would require consumer research. These sensitivity results provide business boundaries that can be used to consider investing in understanding consumers more accurately, e.g., by using larger samples.

Overall, joint optimization of assortment, spacing and pricing in one objective function not only improves the accuracy of the model, but allows the achievement of better objective values as well. The impact of varying consumer behavior and different hierarchical decisions remains medium to moderate.

6.5 Numerical Examples and Test Problems

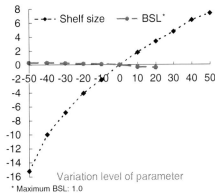
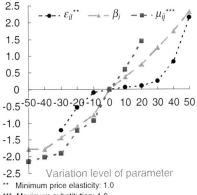

Fig. 6.5 Sensitivity analyses of the CASPP-model

6.5.5 Applicability for Large-Scale Categories

One should note that the scale of the optimization problem depends on the values of three variables: number of product items (I) in the category, number of facing levels (K), and number of price levels (L). The mixed-integer quadratic model features an objective function with linear constraints. In general, mixed-integer quadratic problems (MIQPs) are difficult to solve. However, the case examples show that the computation time remains within reasonable time bounds for the strategic decision problem (≤ 30 min).

The number of variables is $I + I \cdot K \cdot L$, where I denotes the number of items, K the number of discrete facings, and L the number of discrete price levels. Furthermore, using the parameter μ_{ji} reduces the data requirements from $I^{K \cdot L}$ for all assortment, facing and price combinations to a matrix with I^2 combinations.

For further numerical analysis, "large-scale" problem cases are chosen that include very high values for all the three variables consistent with the potential reality of retailers' decision environments. For instance, a large supermarket retailer carries from approximately 35,000–50,000 items within 600 categories, i.e., the average number of distinct items in a category is around 60–80. The largest test is therefore $I = 80$. The variation in facing levels is also restricted to currently existing shelf layout. Retailers prefer to change shelf layout incrementally so as not to confuse consumers. As a result, facing variations are allowed that still fulfill lower and upper bounds. Price image plays a key role in a retailer's strategy. Retailers therefore strategically define the price ranges of categories and products in a long-term decision. This long-term strategy (or these mid-term adjustments) are input to the model presented. Hence, the number of price levels L is also limited, as retailers

Table 6.3 Evaluation of computation performance of the CASPP-model

I	K	L	MIP gap at 100 s	Calc. time for MIP gap < 0.5% (s)
40	5	3	5.4%	212
40	5	5	5.5%	224
40	10	5	5.5%	230
80	5	3	18.2%	434
80	5	5	22.5%	498
80	10	5	30.3%	644

stick to certain price points. $L = 5$ price levels will be allowed for the "large-scale" test.

Table 6.3 summarizes the computation time and the results. The CPLEX solver found a feasible solution where the MIP solution was close to the LP solution. The LP-MIP solution gap with less than 0.5% was achieved after less than 10 minutes. However, CPLEX then took more than 30 minutes to prove that the solution is optimal in all the test cases.

6.6 Conclusions and Future Areas for Research

This chapter is concerned with the development and analysis of an integrated model for assortment size, shelf space management and price optimization. The model developed is a mixed-integer quadratic non-linear problem that was solved using CPLEX.

The model presented extends known shelf space management models using price effects. It is based on consumer decisions and operational constraints. The computation results show that price and space elasticity are efficiently exploited, and that the model avoids high restocking requirements as well as that space is efficiently exploited. It also clarifies restocking requirements to satisfy basic supply levels. This model provides results for retail-specific (sub-) category problem sizes within reasonable computation times. Furthermore, it has been shown that the integrated model provides more accurate and better solutions. It is possible to apply the model to a large set of category optimization problems, as standard retail data was mainly used such as historical sales, which are available at a store level. The sales data only require support via experimental data.

Limitations of the model can be viewed as further areas of research. Some areas relate to (1) hierarchical planning aspects, (2) problem size, and (3) the nature of a deterministic model. First, the restocking costs could be further specified and quantified. Out-of-stock substitution does not occur as efficient instore logistics have been assumed, as in other shelf space models. However, adding costs to restocking processes would mean that it might be more beneficial to save on restocking and allow substitution of other products. The model assumes unlimited

6.6 Conclusions and Future Areas for Research

transportation, warehouse and backroom capacity, as only showroom effects have been integrated. Additionally, positing aspects of a planning hierarchy as input to managerial constraints and enlarging the scope down the supply chain will provide additional insights. This also includes joint shelf price optimization with suppliers and game-theoretic aspects. The further impact of marketing activities and demand effects could be studied, too, e.g., positioning. Moreover, the mixed quadratic problem can only be solved with a limited number of items as in the test cases. Developing and implementing heuristics may overcome this, and the effectiveness of the heuristics can be measured against the model in this chapter. Another possible extension is to consider stochastic demand.

Chapter 7
Conclusions and Outlook

This thesis has provided a coherent retail operations planning architecture for retail demand and supply chain planning. The review of empirical insights, DSSs and commercial software applications shows that category managers require more quantitative analyses for the master planning of retail shelves. Decisions in this area can benefit from a modeling perspective. Shelf space management is one of the most difficult aspects of retailing. A significant reason is that while retail shelf space is fixed, the number of new potential products is constantly growing and evolving alongside consumer desire. Decision support systems (DSS) have therefore been developed at the interface of marketing and operations that integrate assortment planning, price management and inventory management with shelf space management. These models require commonly available retail sales and profit data, and are supported by consumer research data. In that spirit, this thesis aims to bridge the gap between theory and practice and nudge retail category managers towards more quantitative analysis in their shelf space decisions.

7.1 Contribution to Research Status

The concurrence of market trends has made it all the more important for retailers to be as efficient as possible in managing the allocation of their existing shelf space – arguably one of the scarcest and most strategically valuable resources. While retail managers strive to follow the industry mantra "retail is detail," most of them have little time to consider the details of different planning aspects. Therefore a comprehensive planning architecture is required that structures supply and demand planning tasks and arranges them according to hierarchical and vertical interdependencies. DSSs abstract from reality and use models as a basis for plans. Analytical models are emerging as the most promising solutions to many of these planning problems, especially as advances in computing capabilities allow the solution of larger retail problems (Agrawal and Smith 2009b; Kopalle 2010).

The goal has become how best to manage product assortments to generate higher profit from limited shelf space. A key imperative for retailers striving to achieve efficient retail shelves is obviously the availability of appropriate DSSs to help with the more efficient allocation of the shelf space they have available. This study focuses on developing such DSSs.

Models are developed that jointly optimize category managers' decisions on assortment size, number of facings, product prices and restocking requirements in a product category. As a result, the models better capture the realistic same-time decision environment faced by category managers, including several key aspects hitherto treated rudimental by the existing models. The multi-product assortment, shelf space, price and inventory management problems integrate consumer interaction and hierarchical decision aspects. Practical solutions are provided to optimize retail category-specific problem sizes. The numerical examples show the superiority of integrated models over pure successive planning and the rules in commercial planogram-creation software. A fully integrated model that jointly reflects price and inventory considerations has not been provided up to now, as the focus of the model extensions was on analyses of the supply-side and demand-side impact of such aspects.

This research effort proposes models that are accurate, easy to implement, and flexible enough to be applied to a wide range of shelf space management models. The models contribute to the existing literature in four directions. Firstly, unlike existing studies that consider a shelf space allocation decision independent of assortment decisions, the models allow joint decisions on both, and capture cross-product interactions through substitution behavior. Assortment decisions are integrated in shelf space management as out-of-assortment substitution from latent consumer demand. This more realistically reflects consumer instore behavior and category managers' decision problems. Secondly, existing shelf space allocation models consider only "efficient replenishment systems," and abstract from the actual restocking processes of retailers. In contrast, the integrated inventory and shelf space management model considers restocking policies of a fixed scheduled filling before sales begin and concurrent restocking during the sales period. Also, price has been previously treated as an exogenous parameter in shelf space management. The integrated price and shelf space management model relaxes this and uses price adjustments to increase space productivity and category profit. Finally, the models are transformed from non-linear mixed-integer problems to multi-choice knapsack problems. This allows the use of commercial solvers such as CPLEX to obtain a globally optimal shelf space configuration with fast computation times. The numerical examples show the benefit of integrated models on total profit and solution structure. The number of facings, restocking requirements and shelf prices all have significant impact on category profitability. Sensitivity analyses are used to additionally compute error bounds for the parameter estimates. Finally, managerial decisions and constraints on operational fulfillment are analyzed as part of a comprehensive hierarchical retail planning framework.

7.2 Further Areas for Research

Model assumptions and limitations can be viewed as further potential research areas. Four avenues emerge as important future research directions based on the previous discussion in this thesis. It seems that there are significant opportunities in generalizing the existing theoretical work to handle the more complex problems faced by category managers.

First, the planning problems within master category management need to be aligned with hierarchical and vertical planning interdependencies. Second, incorporating other demand effects based on empirical findings on consumer behavior in shelf space optimization models seems a valuable area of research. Third, different modeling techniques and solution procedures may be required to cope with large-scale problems that integrate other demand effects and planning aspects, or that require the integration of stochastic or dynamic effects. Finally, most of the existing theoretical models have not been incorporated into standard applications (meaning their theoretical predictions have not been empirically tested). The field would benefit from such applications and empirical tests as a validation of the assumptions in the increasingly complex shelf space management planning models being formulated in the academic literature. Below some possible research topics from these four avenues are described, in no particular order.

7.2.1 Alignment with Other Planning Problems

7.2.1.1 Vertical Integration of Planning Problems

The model assumes unlimited transportation, warehouse and backroom capacity, as only showroom effects are integrated. Areas of further research are the investigation of joint optimization of space assignment, instore replenishment cycles and order cycles for backroom replenishment. Here, a further constraint in overall replenishment capacity for the scheduled basic fill and product-specific refill processes could also be applied if retailers are constrained where their overall workforce capacity is concerned. It would also be worthwhile to study the impact of case pack sizes and alignment using direct filling of shelves with warehouse deliveries (without backroom storage). Backroom capacity and backroom inventory costs could also be integrated, as only showroom effects and costs have been reflected so far.

7.2.1.2 Hierarchical Integration of Planning Problems

The shelf space management models are capable of allocating shelf space to products within a product category. The models use a bottom-up approach for shelf space assignment to individual products and implicitly assume that the amount of shelf space assigned to a product category is predetermined (Irion et al. 2004). The space allocation models, which assign shelf space to individual products, are flexible enough to be applied to allocation problems at a higher level of aggregation. Thus, similar models can be used to allocate store shelf space to entire categories using a top-down approach. In a hierarchical framework, the results of category space assignment become a constraint for product space assignment within a category, and would this need to communicate back to the overarching layout configuration. However, these two different hierarchical approaches have not yet been connected. This research area appears very promising, as sensitivity analyses across the models indicate the high impact of total shelf space on profit. Also, further master category planning activities like category sales planning and sales forecasting should be included in the analyses of overarching planning requirements.

Anticipations of the subordinated instore logistics planning should also be reflected in the decision model. Retailers are constrained in their shelf replenishment due to limitations on the shelf merchandizers available to immediately fill the shelves after stock out and the expensive handling costs within stores. The further integration of replenishment constraints and non-linear shelf stacking costs seems a worthwhile area of research.

7.2.2 Further Demand Effects

Consumer purchase decisions across product categories may not always be independent. Further, while in this research the objective was to optimize category profit, other performance criteria exist that need be to evaluated, such as market basket or cross-category metrics (Cachon et al. 2005; Cachon and Kök 2007; Kamakura and Wooseong 2007). Since shelf space management often interacts with other marketing mix activities, future research is needed that investigates the effect of cross-category effects, price promotions and advertising (Hansen et al. 2010). Wickern (1966, p.41) explain that "the success of retailing consists not only of selling merchandise, but also of the nature and completeness of the assortment."

Shopping is a hedonistic event for a growing number of customers (e.g., Arnold and Reynolds 2003). Customers are increasingly affected by store layout and susceptible to impulse buying (e.g., Kaltcheva and Weitz 2006; Massara and Pelloso 2006; Mattila and Wirtz 2008; Liu et al. 2007). Consequently, category managers often place products together according to brands or product sizes to create a category image. Pure shelf space profit maximization could work against such imagery. Research is lacking that balances shelf space assignment and atmospherics. As Hansen et al. (2010, p.102) concludes, that until the effects of visual layout effects are not integrated, "practitioners would be wise to also keep a 'human touch' in the planogram design process."

Further marketing activities and demand-generating effects should be investigated. These include, in particular, positioning effects to account for different shelf layers and "eye-level" demand, the impact of promotional effects on permanent assortment, and other marketing variables that generate instore demand.

Competitive scenarios that additionally influence overall store sales could also be part of an integrated analysis.

Another shortcoming in the modeling is the implicit assumption that shelves are always kept fully stocked. While some existing models allow partial shelf depletion and calculate this impact on demand (Urban 1998; Hariga et al. 2007), the effect of consumer behavior on out-of-stock substitution has not been investigated in these models. Out-of-stock substitution does not occur as efficient instore logistics has been assumed. However, integrating out-of-stock substitution and restocking costs in space assignment could lead to a situation where it might be more beneficial to save on stocking, instead allowing substitution of other products.

7.2.3 Empirical Validation of Model Recommendations

More empirical analyses are needed to understand the impact of merchandizing variables on consumer choice and purchasing behavior. Only limited empirical studies have been conducted of the relationship of assortment, shelf space and inventory decisions with other levers such as pricing, promotions, and advertising (e.g., McIntyre and Miller 1999; Richards and Hamilton 2006). Joint optimization of some of these variables may lead to interesting results.

Estimating model parameters such as substitution probabilities and space elasticity is another area that needs further research. There is an extensive body of literature in marketing and econometrics that deals with estimating parameter for a wide variety of consumer choice models (Kök et al. 2009). However, there is little application of these in the shelf space planning literature. To bridge retail practice and academic research, it is important to come up with innovative and cost-efficient techniques to estimate the parameters that form the backbone of several optimization models.

7.2.4 Modeling Techniques

The computing power of machines and functionality of solvers have gone up exponentially. Data storage restrictions have also declined at a similar pace. These changes, combined with quantitatively trained retail managers with an analytics orientation at many organizations, have provided a great opportunity for modelers in retailing research to actually have a practical impact (Kopalle 2010). However, even though it has been shown that the models presented are capable of providing solutions for practical category sizes, the integration of further demand effects and planning aspects may require the use of specialized heuristics or metaheuristics.

Furthermore, consumer demand is assumed to be deterministically known, but in fact is subject to certain volatility depending on external factors like season, temperature or weekday. An extension to stochastic and non-stationary demand would further improve the reliability of results.

Finally, scientific models take a static view of the shelf space planning problem, whereas in practice, shelf space decisions in a category may be made several times throughout the season. The dynamic shelf space problem provides a rich set of research questions, such as shelf space planning with demand learning using tests in sample stores.

7.2.5 Transfer to Commercial Software Applications and Retail Practice

The complexity of managing shelf space will still require being able to visualize of the planogram for final human verification. Any decision support model therefore also needs to communicate with and be adjusted to shelf space software programs, which will be used as visual templates.

Shelf space planning in multi-format, multi-store and multi-category retail chains is a completely open research area. The pros and cons of centralized and hierarchical planning and the execution problems associated with this have not been studied empirically or analytically. Retailers increasingly see balancing the benefits of customizing planograms by store with the increased cost of complexity as a significant source of competitive advantage. A factor possibly counting against this, however, is that strategic corporate plans may be against store managers' "gut feeling" and experience, and result in imperfect execution. An interesting research question here is how to manage the trade-off between "one size fits all" and "each store to its own." The incentive conflict between central retail planners, store managers and consumer goods producers is also a valuable area of research.

Advances in computing resources have permitted the development of more complex shelf space models that are more consistent with consumer decision making. Category managers can use these shelf space models to improve their decision making. However, practitioners tend to use simplistic software tools that can handle large-scale assortments. Science and practice show a big discrepancy in this field. Academic models have greater analytical capabilities than the commercial software available, yet they still need to prove implementability. Science focuses on the approach of large-scale integration of extensive interdependencies of demand, resulting in complicated and expensive requirements for estimating parameters. The main barriers to retailer adoption (and thus areas of investigation) are requirements for large item sets, dynamically changing consumer preferences, model complexity, difficulty of integration into existing systems, and interfaces between marketing and operations.

This research extended the existing literature that addresses the shelf space allocation problem. It does so by capturing the critical decision trade-offs faced by retailers in optimizing their shelf space. This dissertation structures the planning problems, devises decision support models to maximize category profit, and provides methods to test the capability of models for category-specific problems. The planning issues are illustrated using models with differing levels of integration.

Specifically, this dissertation develops models that optimize retailers' decisions relating to assortment size, number of facings, replenishment frequency and product prices in a retail category. It would be hugely rewarding to see these models integrated into day-to-day retail practice, with all the benefits that this would imply for both the trade and ultimately for the customer.

Bibliography

Abbott H, Palekar US (2008) Retail replenishment models with display-space elastic demand. European J Oper Res 186(2):586–607

Abraham MM, Lodish LM (1993) An implemented system for improving promotion productivity using store scanner data. Market Sci 12(3):248–269

Adenso-Díaz B, González M, García E (1998) A hierarchical approach to managing dairy routing. Interfac 28(2):21–31

Agrawal N, Smith SA (2009a). Mulit-location inventory models for retail supply chain management. In: Agrawal N, Smith SA (ed) Retail supply chain management. International Series in Operations Research & Management Science. Springer, Boston, MA

Agrawal N, Smith SA (eds) (2009b). Retail supply chain management: Quantitative models and empirical studies. International Series in Operations Research & Management Science. Springer, Boston, MA

Agrawal N, Smith SA (2009c). Supply chain planning process for two major retailers. In: Agrawal N, Smith SA (eds) Retail supply chain management. International Series in Operations Research & Management Science. Springer, Boston, MA

Agrawal N, Smith SA, Tsay AA (2002) Multi-vendor sourcing in a retail supply chain. Prod Oper Manag 11(2):158–182

Akkerman R, Farahani P, Grunow M (2010) Quality, safety and sustainability in food distribution: A review of quantitative operations management approaches and challenges. OR Spectrum 32(4):863–904

Alvarado UY, Kotzab H (2001) Supply chain management: The integration of logistics in marketing. Ind Market Manag 30(2):183–198

Angerer A (2006) The impact of automatic store replenishment on retail: Technologies and concepts for the out-of-stocks problem. Gabler, Wiesbaden

Anily S, Bramel J (1999) Vehilce routing and the supply chain. In: Tayur S, Ganeshan R, Magazine M (ed) Quantitative models for supply chain management. Kluwer Academic Publishers Group, Dordrecht, pp. 149–196.

Annupindi R, Dada M, Gupta S (1997) A dynamic model of consumer demand with stock out based substitution. Ph.D. thesis, Kellog School of Management, Evanston, IL

Anupindi R, Gupta S, Venkatarmanan MA (2009) Managing variety on the retail shelf: Using household scanner panel data to rationalize assortments. In: Agrawal N, Smith SA (ed) Retail supply chain management. International Series in Operations Research & Management Science. Springer, Boston, MA

Arnold MJ, Reynolds KE (2003) Hedonic shopping motivations. J Retailing 79(2):77–95

Aviv Y (2001) The effect of collaborative forecasting on supply chain performance. Manag Sci 47(10):1326–1343

Aviv Y (2007) On the benefits of collaborative forecasting partnerships between retailers and manufacturers. Manag Sci 53(5):777–794

Aydin G, Porteu EL (2009) Manufacturer-to-retailer versus manufacturer-to-consumer rebates in a supply chain. In: Agrawal N, Smith SA (ed) Retail supply chain management. International Series in Operations Research & Management Science. Springer, Boston, MA

Basuroy S, Mantrala MK, Walters RG (2001) The impact of category management on retailer prices and performance: Theory and evidence. J Marke 65(October):16–32

Becker J, Uhr W, Vering O (2000) Integrierte Informationssysteme in Handelsunternehmen auf der Basis von SAP. Springer, Berlin

Bertrand JW, Fransoo JC (2002) Operations management research methodologies using quantitative modelling. Int J Oper Prod Manag 22(2):241–264

Bhattacharjee S, Ramesh R (2000) A multi-period profit maximizing model for retail supply chain management: An integration of demand and supply-side mechanisms. Eur J Oper Res 122:584–601

Bimolt TH, van Heerde HJ, Pieters RG (2005) New empirical generalizations on the determinants of price elasticity. J Market Res 38(May):141–156

Binkley JK, Connor JM (1998) Grocery market pricing and the new competitive environment. J Retailing 74(2):273–294

Bish E, Maddah B (2004) On the interaction between variety, pricing, and inventory decisions under consumer choice: Working Paper. Technical Report, 1–33

Bitran GR, Caldentey R (2003) An overview of pricing models for revenue management. Manuf Ser Oper Manag 3(5):203–229

Blackwell RD, Blackwell K (1999) The century of the consumer: Converting supply chains into demand chains. Supply Chain Manag Rev 3(Fall):22–32

Blattberg RC, Neslin SA (1993) Sales promotion models. In: Eliashberg J, Lillien GL (ed) Marketing. Elsevier, Amsterdam: North-Holland, pp. 553–609

Boatwright P, Nunes JC (2001) Reducing assortment: An attribute-based approach. J Marke 65: 50–63

Borin N, Farris P (1995) A sensitivity analysis of retailer shelf management models. J Retailing 71:153–171

Borin N, Farris P, Freeland J (1994) A model for determining retail product category assortment and shelf space allocation. Decis Sci 25(3):359–384

Bozer YA, Kile JW (2008) Order batching in walk-and-pick order picking systems. Int J Prod Res 46(7):1887–1909

Breuer P, von Fritsch A, Prauschke C, Steegmann J (2009) Auf der Suche nach dem richtigen Angebot: Sortiment und Kundenverständnis sind Top-Themen im Krisenjahr. Akzente 4(3): 8–13

Broekmeulen R, van Donselaar K, Fransoo J, van Woensel T (2006) The opportunity of excess shelf space in grocery retail stores. Technical Report, 1–34

Broniarczyk SM, Hoyer WD, McAlister L (1998) Consumers' perceptions of the assortment offered in a grocery category: The impact of item reduction. J Market Res 35(2):166–176

Brown MG, Lee JY (1996) Allocation of shelf space: A case study of refrigerated juice products in grocery stores. Agribusiness 12(2):113–121

Brown WM, Tucker WT (1961) The marketing center: Vanishing shelf space. Atlanta Econ Rev 9:9–13

Bucklin RE, Russell GJ, Srinivasan V (1998) A relationship between market share elasticities and brand switching probabilities. J Market Res 35(February):99–113

Bultez AP, Naert P, Gijsbrechts E, Abelle PV (1989) Asymmetric cannibalism in retail assortments. J Retailing 65(2):153–192

Bunn DW, Vassilopoulos AI (1999) Comparison of seasonal estimation methods in multi-item short-term forecasting. Int J Forecast 15:431–443

Büschken J (2009) Conversion of shoppers in brick-and-mortar retailing: An analysis of observational data. SSRN, 1–38. URL http://ssrn.com/abstract=1394241

Cachon G, Terwiesch C, Xu Y (2005) Retail assortment planning in the presence of consumer search. Manuf Serv Oper Manag 7(4):330–346

Cachon GP, Kök GA (2007) Category management and coordination in retail assortment planning in the presence of basket shopping consumers. Manag Sci 53(6):934–951

Campillo-Lundbeck S (2009) Die Regeln im Regal: Category Management. Unternehmen & Märkte (06), 30–34

Campo K, Gijsbrechts E (2005) Retail assortment, shelf and stockout management: Issues, interplay and future challenges. Appl Stoch Model Bus Ind 21(4-5):383–392

Campo K, Gijsbrechts E, Goossens T, Verhetsel A (2000) The impact of location factors on the attractiveness and optimal space shares of product categories. Int J Res Marketing 17(4):255–279

Campo K, Gijsbrechts E, Nisol P (2003) The impact of retailer stockouts on whether, how much, and what to buy. Int J Res Marketing 20(3):273–286

Cannondale Associates (2003) Category management: 2003 Benchmarking study. Cannondale Associates Inc., Evanston, IL

Cardós M, García-Sabater JP (2006) Designing a consumer products retail chain inventory replenishment policy with the consideration of transportation costs: Theoretical issues in production scheduling and control & planning and control of supply chains and production. Int J Prod Econ 104(2):525–535

Carlotti Jr S, Coe ME, Perrey J (2006) Designing and managing winning brand portfolios: Profiting from proliferation. McKinsey Quarterly

Chandon P, Hutchinson WJ, Bradlow ET, Young SH (2009) Does in-store marketing work? Effects of the number and position of shelf facings on brand attention and evaluation at the point of purchase. J Market 73(November):1–17

Christiani P, Küpper J, Sänger F, Theunissen R (2009) Weniger ist mehr: So optimieren Hersteller ihr Portfolio: Komplexe Sortimente sind teuer. Mit weniger Produkten zu mehr Umsatz. Akzente 3(2):38–43

Corsten DS, Gruen TW (2003) Desperately seeking shelf availability: An examination of the extent, the causes, and the efforts to address retail out-of-stocks 31(12):605–617

Corstjens M, Doyle P (1981) A model for optimizing retail space allocations. Manag Sci 27(7):822–833

Curhan RC (1972) The relationship between shelf space and unit sales in supermarkets. J Market Res 9:406–412

Curseu A, van Woensel T, Fransoo J, van Donselaar K, Broekmeulen R (2009) Modelling handling operations in grocery retail stores: An empirical analysis. J Oper Res Soc 60(2):200–214

Dallari F, Marchet G, Melacini M (2008) Design of order picking system. Int J Adv Manuf Tech 42:1–12

De Koster MBM, Van der Poort ES, Wolters M (1999) Effcient orderbatching methods in warehouses. Int J Prod Res 37(7):1479–1504

de Koster RBM, Le-Duc T, Roodbergern KJ (2007) Design and control of warehouse order picking: A literature review. Eur J Oper Res 182(2):481–501

DeHoratius N, Ton Z (2009) The role of execution in managing product availability. In: Agrawal N, Smith SA (ed) Retail supply chain management. International Series in Operations Research & Management Science. Springer, Boston, MA

Dekker M, van Donselaar K, Ouwehand P (2004) How to use aggregation and combined forecasting to improve seasonal demand forecasts. Int J Prod Econ 90:151–167

Desmet P, Renaudin V (1998) Estimation of product category sales responsiveness to allocated shelf space. Int J Res Market 15(5):443–457

Desrochers DM, Nelson P (2006) Adding consumer behavior insights to category management: Improving item placement decisions. J Retailing 82(4):357–365

Dhar SK, Hoch SJ, Kumar N (2001) Effective category management depends on the role of the category. J Retailing 77:165–184

Drèze X, Hoch SJ, Purk ME (1994) Shelf management and space elasticity. J Retailing 70(4): 301–326

Drezner T (2009) Location of retail facilities under conditions of uncertainty. Ann Oper Res 167(1):107–120
Durvasula S, Sharma S, Andrews CJ (1992) STORELOC: A retail store location model based on managerial judgements. J Retailing 68(Winter):420–444
ECR Europe (2003a). Handbook ECR-Demand Side. EHI, Cologne
ECR Europe (2003b). Optimal shelf availability: Increasing shopper satisfaction at the moment of truth. URL http://www.ecrnet.org/04-publications/blue_books/pub_2003_osa_blue_book.pdf
Efficient Consumer Response (1993) Enhancing consumer value in the grocery industry. Food Marketing Institute, Washingtion, DC
EHI Retail Institute (2010) Handel aktuell 2009/2010: Struktur, Kennzahlen und Profile des internationalen Handels - Schwerpunkt Deutschland, Österreich, Schweiz
Emmelhainz LW, Emmelhainz MA, Stock JR (1991) Logistics implications of retail stockouts. J Bus Logist 12(2):129–142
Erlebacher SJ, Meller RD (2000) The interaction of location and inventory in designing distribution systems. IIE Tran 32:155–166
Federgruen A, Heching A (1999) Combined pricing and inventory control under uncertainty. Oper Res 47:454–477
Fernie J (1999) Outsourcing distribution in U.K. retailing. J Bus Logis 20(2):83–96
Fernie J, Sparks L (2009) Logistics and retail management: Insights into current practice and trends from leading experts, 2nd Edition. Clays, London
Fernie J, Staines H (2001) Towards an understanding of European grocery supply chains. J Retailing Consum Serv 8:29–36
Fisher ML, Raman A (2010) The new science of retailing: How analytics are transforming the supply chain and improving performance. Harvard Business School Publishing, Boston.
Fitzsimons GJ (2000) Consumer response to stockouts. J Consum Res 27:249–266
Fleischmann B, Meyr H (2003) Planning hierarchy, modeling and advanced planning systems: Chapter 9. In: Graves SC, de Kok AG (ed) Supply chain management: Design, coordination, operation. Elsevier, Amsterdam, pp. 457–523
Fleischmann B, Meyr H, Wagner S (2008) Advanced planning. In: Stadtler H, Kilger C (ed) Supply chain management and advanced planning. Springer, Berlin, pp. 81–106
Friend SC, Walker PH (2001) Welcome to the new world of merchandising. Harvard Business Review (November) 133–141
Gajjar HK, Adil GK (2008) A piecewise linearization for retail shelf space allocation problem and a local search heuristic. Ann Oper Res 1–19
Gajjar HK, Adil GK (2011) Heuristics for retail shelf space allocation problem with linear profit. Int J Retail Distrib Manag 39(2):144–145
Gallego G, van Ryzin GJ (1994) Optimal dynamic pricing of inventories with stochastic demand over finite horizons. Manag Sci 40(8):999–1020
Ganeshan R (1999) Managing supply chain inventories: A multiple retailer, one warehouse, multiple supplier model. Int J Prod Econ 59:341–354
Ganeshan R, Ring LJ, Strong JS (2007) A mathematical approach to designing central vs. local ordering in retail. Int J Prod Econ 108:341–348
Gaur V, Fisher ML (2005) In-store experiments to determine the impact of price on sales. Prod Oper Manag 14(4):377–387
Gaur V, Honhon D (2006) Assortment planning and inventory decisions under a locational choice model. Manag Sci 52:1528–1543
Gebhardt M, Kuhn H (2008) Hierarchische Produktions- und Logistikplanung bei unvollkommener Information. ZfB-Sonderband Logistik (4):99–124
Gill A, Bhatti MI (2007) Optimal model for warehouse location and retailer allocation. Appl Stoch Model Bus Ind 23:213–221
Granzin KL, Painter JJ, Valentin EK (1997) Consumer logistics as a basis for segmenting retail markets: An exploratory inquiry. J Retailing and Consum Serv 4(2):99–107
Greenhouse S (2005) How Costco became the Anti-Wal-Mart. New York Times
Grewal D, Levy M (2007) Retailing research: Past, present, and future. J Retailing 83(4):447–464

Griswold M (2007) Space management: Align business challenges and IT vendors. AMR Res 1–17

Grocery Manufacturers Association, AC Nielsen, McKinsey&Company (2005) Winning with customers to drive real results: The 2005 customer and channel management survey. Grocery Manufacturers Association, Washington, D.C

Groothedde B, Rujigrok C, Tavasszy L (2005) Towards collaborative, intermodal hub networks: A case study in the fast moving consumer goods market. Transport Res Part E 41(6):567–583

Gruen TW, Shah RH (2000) Determinants and outcomes of plan objectivity and implementation in category management relationships. J Retailing 76(4):483–510

Gruen WT, Corsten S, Bharadwaj S (2002) Retail out-of-stocks: A worldwide examination of extent, causes and consumer responses: Report. Grocery Manufacturers of America, Washingtion, D.C

Gu J, Goetschalckx M, McGinnis LF (2007) Research on warehouse operation: A comprehensive review. Eur J Oper Res 177(1):1–21

Günther H-O, Meyr H (2009) Supply Chain Planning: Quantitative Decision Support and Advanced Planning Solutions. Springer, Berlin

Gupta S (1988) Impact of sales promotions on when, what, and how much to buy. J Market Res 25(4):342–355

Gutgeld Y, Sauer S, Wachinger T (2009) Wachstum - aber wie? Akzente 3(3):14–19

Hadley G, Whitin TM (1963) Analysis of Inventory Systems. Prentice-Hall, Englewood Cliffs, NJ

Hall JM, Kopalle PK, Krishna A (2010) Retailer dynamic pricing and ordering decisions: category management versus brand-by-brand approaches: Special issue: Modeling retail phenomena. J Retailing 86(2)172–183

Hamister JW, Suresh NC (2008) The impact of pricing policy on sales variability in a supermarket retail context: Special section on sustainable supply chain. Int J Prod Econ 111(2)441–455

Hansen JM, Raut S, Swami S (2010) Retail shelf allocation: A comparative analysis of heuristic and meta-heuristic approaches. J Retailing 86(1):94–105

Hansen P, Heinsbroek H (1979) Product selection and space allocation in supermarkets. Eur J Oper Res 3(6):474–484

Hariga MA, Al-Ahmari A, Mohamed A-RA (2007) A joint optimisation model for inventory replenishment, product assortment, shelf space and display area allocation decisions. Eur J Oper Res 181(1):239–251

Helm R, Stölzle W (2009) Optimal Shelf Availability: Effiziente Managementkonzepte zur Optimierung der Regalverfügbarkeit. Deutscher Fachverlag, Frankfurt am Main

Helnerus K (2009) Die Lücke im Regal: Out-of-Stock Situationen aus Sicht der Kunden und des Handelsmanagements. Kohlhammer, Stuttgart

Helnerus K, Müller-Hagedorn L (2006) Die Lücke im Regal: Out-of-Stock-Situationen aus Sicht der Kunden und des Handelsmanagements. Handel im Fokus 58(4):212–224

Hennessy T (2001). Categorizing Sucees. Progressive Grocer 80(4)

Hernandez T, Bennison D, Cornelius S (1998) The organisational context of retail locational planning. GeoJournal 45(4):299–308

Hertel J, Zentes J, Schramm-Klein H (2005) Supply-Chain-Management und Warenwirtschaftssysteme im Handel. Springer, Berlin

Hoch SJ, Deighton JA (1989) Managing what consumers learn from experience. J Market 53(2):1–20

Hoch SJ, Drèze X, Purk ME (1994) EDLP, Hi-Lo and margin arithmetic. J Market 58:16–27

Hopp W, Xu X (2008) A static approximation for dynamic demand substitution with applications in a competitive market. Oper Res 56:630–645

Hoyer WD (1984) An examination of consumer decision making for a common repeat purchase product. J Consum Res 11(3):822–831

Hübner AH, Kuhn H (2011a). Assortment, shelf space and inventory management with space-elastic demand and substitution effects. Technical Report KU Eichstätt-Ingolstadt, 1–25

Hübner AH, Kuhn H (2011b). Integrative retail assortment, shelf space and price planning. Technical Report KU Eichstätt-Ingolstadt, 1–27

Hübner AH, Kuhn H (2011d). Retail shelf space management model with space-elastic demand and consumer-driven substitution effects. Technical Report KU Eichstätt-Ingolstadt, 1–19. URL http://ssrn.com/abstract=1534665

Hübner AH, Kuhn H, (2011e). Shelf and inventory management with space-elastic demand. In: Hu B, Morasch K, Siegle M, Pickl S (eds) Operations Research Proceedings 2010. Springer, Berlin, p. (in press)

Hbner AH, Kuhn H (2012). Retail category management: State-of-the-art review of quantitative research and software applications in assortment and shelf space management. Omega 40(2):199–209

Hübner AH, Kuhn H, Sternbeck MG (2010) Demand and supply chan planning in grocery retail: An operations planning framework. Technical Report KU Eichstätt-Ingolstadt, 1–29 URL http://ssrn.com/abstract=1635752

Hui SK, Bradlow ET, Fader PS (2009) Testing behavioral hypotheses using an integrated model of grocery store shopping path and purchase behavior. J Consum Res 36:478–493

Hwang H, Choi B, Lee M-J (2005) A model for shelf space allocation and inventory control considering location and inventory level effects on demand. Int J Prod Econ 97(2):185–195

Hwang H, Hahn KH (2000) An optimal procurement policy for items with an inventory level-dependent demand rate and fixed lifetime. Eur J Oper Res 127(3):537–545

Hwang H, Oh YH, Lee YK (2004) An evaluation of routing policies for order-picking operations in low-level picker-to-part system. Int J Prod Res 14(18):3873–3889

Inman JJ, Winter RS, Ferraro R (2009) The interplay among category characteristics, customer characteristics, and customer activities on in-store decision making. J Market 73(September):19–29

Irion J, Al-Khayyal F, Lu J (2004) A piecewise linearization framework for retail shelf space management models: Working Paper. Technical Report, 1–25

Iyengar P, Lepper M (2000) When choice is demotivating: Can one desire too much of a good thing? J Pers Soc Psychol 79(6):995–1006

Iyer ES (1989) Unplanned purchasing: Knowledge of shopping environment and time pressure. J Retailing 65(1):40–57

Jayaraman V, Patterson RA, Rolland E (2003) The design of reverse distribution networks: Models and solution procedures. Eur J Oper Res 150:128–149

Kabadayi S, Eyuboglu N, Thomas GP (2007) The performance implications of designing multiple channels to fit with strategy and environment. J Market 71(October):195–211d

Kabak O, Ülengin F, Aktas E, Önsel S, Topcu YI (2008) Efficient shift scheduling in the retail sector through two-stage optimization. Eur J Oper Res 184:76–90

Kaltcheva VD, Weitz BA (2006) When should a retailer create an exciting store environment? J Mark 70:107–118

Kamakura W, Wooseong K (2007) Chain-wide and store-level analysis for cross-category management. J Retailing 83(2):159–170

Kambil A, Agrawal V (2001) The new realities of dynamic pricing. Accenture Outlook J (2):15–21

Kellerer H, Pferschy U, Pisinger D (2004) Knapsack problems. Springer, Berlin

Ketzenberg M, Metters R, Vargas V (2002) Quantifying the benefits of breaking bulk in retail operations. Int J Prod Econ 80(3):249–263

Khouja M (1999) The single-period (news-vendor) problem: Literature review and suggestions for future research. Omega 27(5):537–553

Kök GA (2003) Management of product variety in retail operations: Ph.D. Dissertation. The Wharton School, University of Pennsylvania, Pennsylvania

Kök GA, Fisher ML (2007) Demand estimation and assortment optimization under substitution: Methodology and application. Oper Res 55(6):1001–1021

Kök GA, Fisher ML, Vaidyanathan R (2009) Assortment planning: Review of literature and industry practice. In: Agrawal N, Smith SA (ed) Retail supply chain management. International Series in Operations Research & Management Science. Springer, Boston, MA, pp. 99–154

Kopalle P, Biswas D, Chintagunta PK, Fan J, Pauwels K, Ratchford BT, Sills JA (2009) Retailer pricing and competitive effects: Enhancing the retail customer experience. J Retailing 85(1): 56–70

Kopalle PK (2010) Special issue: Modeling retail phenomena. J Retailing 86(2):117–124

Koschat MA (2008) Store inventory can affect demand? Empirical evidence from magazine retailing. J Retailing 84(2):165–179

Kotzab H (1999) Improving supply chain performance by efficient consumer response? A critical comparison of existing ECR approaches. J Bus Ind Market 14(5):364–377

Kotzab H, Teller C (2005) Development and empirical test of a grocery retail instore logistics model. Br Food J 107(8):594–605

Kuhn H, Sternbeck MG (2011) Logistik im Lebensmittelhandel: Eine empirische Untersuchung zur Ausgestaltung handelsinterner Liefernetzwerke. Forschungsbericht der Wirtschaftswissenschaftlichen Fakultät Ingolstadt, Ingolstadt

Kumar M, Patel NR (2008) Using clustering to improve sales forecasts in retail merchandising. URL http://ieweb.uta.edu/vchen/AIDM/AIDM-Kumar.pdf

Kumar S (2008) A study of the supermarket industry and its growing logistics capabilities. Int J Retail Distrib Manag 36(3):192–211

Kurtulus M, Toktay BL (2009) Category captainship practices in the retail industry. In: Agrawal N, Smith SA (ed) Retail supply chain management. International Series in Operations Research & Management Science. Springer, Boston, MA

Kurtulus M, Toktay BL (2011) Category captainship vs. retailer category management and limited retail shelf space. Prod Oper Manag 20(1):47–56

Lam S, Vandenbosch M, Pearce M (1998) Retail sales force scheduling based on store traffic forecasting. J Retailing 74(1):61–88

Lam SY (2001) The effects of store environment on shopping behaviours: A critical review. Adv Consum Res 28:190–197

Le Blanc HM, Cruijssen F, Fleuren HA, De Koster MBM (2006) Factory gate pricing: An analysis of the Dutch retail distribution. Eur J Oper Res 174(3):1950–1967

Le-Duc T, de Koster RBM (2007) Travel time estimation and order batching in a 2-block warehouse. Eur J Oper Res 176(1):374–388

Lehnert M, Hüffner G (2006) IT-gestütztes Category-Management: Status quo und Entwicklungstendenzen. In: Zentes J (ed) Handbuch Handel. Gabler, Wiesbaden, pp. 943–962

Levy M, Grewal D (2000) Supply chain management in a networked economy. J Retailing 76(4):415–429

Levy M, Grewal D, Kopalle PK, Hess JD (2004) Emerging trends in retail pricing practice: implications for research. J Retailing 80(3):xiii–xxi

Levy M, Grewala D (2007) Retailing research: Past, present, and future. J Retailing 83(4):447–464

Li Z (2007) A single-period assortment optimization model. Production and Oper Manag 16(3):369–380

Lim A, Rodrigues B, Zhang X (2004) Metaheuristics with local search techniques for retail shelf-space optimization. Manag Sci 50(1):117–131

Liu S, Melara R, Arangarasan R (2007) The effects of store layout on consumer buying behavioral parameters with visual technology. J Shopping Center Res 14(2):63–72

Lowson R (2001) Analysing the effectiveness of european retail sourcing strategies. Eur Manag J 19(5):534–551

Lyu J, Ding JH, Chen PS (2010) Coordination replenishment mechanisms in supply chain: From the collaborative supplier and store-retail perspective. Int J Prod Econ 123:221–234

Mahajan S, van Ryzin G (2001) Stocking retail assortments under dynamic consumer substitution. Oper Res 49:334–351

Mantrala MK, Levy M, Kahn BE, Fox EJ, Gaidarev P, Dankworth B, Shah D (2009) Why is assortment planning so difficult for retailers? A framework and research agenda. J Retailing 85(1):71–83

Martín-Herrán G, Taboubi S, Zaccour G (2006) The impact of manufactures' wholesale prices on a retailer's shelf-space and pricing decisions. Decis Sci 37(1):71–90

Martínez-de Albéniz V, Roels G (2011) Competing for shelf space. Prod Oper Manag 20(1):32–46

Massara F, Pelloso G (2006) Investigating the consumer–environment interaction through image modelling technologies. Int Rev Retail Distrib Consum Res 16(5):519–531

Mattila AS, Wirtz J (2008) The role of store environmental stimulation and social factors on impulse purchasing. J Serv Market 22(7):562–567

McIntyre SH, Miller CM (1999) The selection and pricing of retail assortments: An empirical approach. J Retailing 75(3):295–318

McKinnon AC, Mendes D, Nababteh M (2007) In-store logistics: An analysis of on-shelf availability and stockout responses for three product groups. Int J Logis 10(3):251–268

Mendes AB, Themido IH (2004) Multi-outlet retail site location assessment. Int Trans Oper Res 11(1):1–18

Meyr H (2004) Supply chain planning in the German automotive industry. OR Spectrum 26(4):447–470

Meyr H, Stadtler H (2008) Types of supply chains. In: Stadtler H, Kilger C (ed) Supply chain management and advanced planning. Springer, Berlin, pp. 65–80

Miller T (2001) Hierarchical operations and supply chain planning. Springer, Berlin

Minner S, Transchel S (2010) Periodic review inventory-control for perishable products under service-level constraints. OR Spectrum 32(4):979–996

Murray CC, Talukdar D, Gosavi A (2010) Joint optimization of product price, display orientation and shelf-space allocation in retail category management: Special Issue: Modeling Retail Phenomena. J Retailing 86(2):125–136

Netessine S, Rudi N (2003) Centralized and competitive inventory models with demand substitution. Oper Res 51(2):329–335

Nielsen AC (2004) Consumer-centric category management: How to increase profits by managing categories based on consumer needs. Wiley, New Jersey

Pal JW, Byrom JW (2003) The five 5s of retail operations: A model and tool for improvement. Int J Retail Distrib Manag 31(10):518–528

Pamuk S, Köksalan M, Güllü R (2004) Analysis and improvement of the product delivery system of a beer producer in Ankara. J Oper Res Soc 55(11):1137–1144

Pentico D (2008) The assortment problem: A survey. Eur J Oper Res 190(2):295–309

Pisinger D (1995) An expanding-core algorithm for the exact 0-1 knapsack problem. Eur J Oper Res 87(1):175–187

Pisinger D (1999) An exact algorithm for large multiple knapsack problems. Eur J Oper Res 114(3):528–541

Pisinger D (2005) Where are the hard knapsack problems? Comput Oper Res 32(9):2271–2284

Potter A, Mason R, Lalwani C (2007) Analysis of factory gate pricing in the UK grocery supply chain. Int J Retail Distrib Manag 35(10):821–834

Quante R, Meyr H, Fleischmann M (2009) Revenue management and demand fulfillment: Matching applications, models and software. OR Spectrum 31(1):31–62

Rajaram K, Tang CS (2001) The impact of product substitution on retail merchandising. Eur J Oper Res 135(3):582–601

Ramaseshan B, Achuthan NR, Collinson R (2008) Decision support tool for retail shelf space optimization. Int J Inform Tech Decis Making 7(3):547–565

Ramaseshan B, Achuthan NR, Collinson R (2009) A retail category management model integrating shelf space and inventory levels. Asia-Pacific J Oper Res 26(4):457–478

Retail Information Systems, Gartner, Oracle (2009) 19th Annual retail technology study: Innovate, execute, adapt. Edgell Communications, NJ

Richards TJ, Hamilton SF (2006) Rivalry in price and variety among supermarket retailers. Am J Agr Econ 88(3):710

Roodbergen KJ, Sharp GP, Vis IFA (2008) Designing the layout structure of manual order picking areas in warehouses. IIE Trans 40(11):1032–1045

Schary PB, Christopher M (1979) The anatomy of a stock-out. J Retailing 55(2):59–70

Schneeweiss C (1998) Hierarchical planning in organizations: Elements of a general theory: Production economics: The link between technology and management. Int J Prod Econ 56–57(9):547–556

Schneeweiss C (2003a). Distributed decision making, 2nd Edition. Springer, Berlin

Schneeweiss C (2003b). Distributed decision making: A unified approach. Eur J Oper Res 150(2):237–252

Schramm-Klein H, Morschett D (2004) Effektivitäts- und Effizienzsteigerungenpotenziale durch IT im Handel. In: Trommsdorff, V. (Ed.), Handelsforschung. Kohlhammer

Shah J, Avittathur B (2007) The retailer multi-item inventory problem with demand cannibalization and substitution: Special section on contextualisation of supply chain networks: 9th International Symposium in Logistics. Int J Prod Econ 106(1):104–114

Shugan SM, Desiraju R (2001) Retail product-line pricing strategy when costs and products change. J Retailing 77(1):17–38

Sloot LM, Verhoef PC (2008) The impact of brand delisting on store switching and brand switching intentions. J Retailing 84(3):281–296

Sloot LM, Verhoef PC, Franses PH (2005) The impact of brand equity and the hedonic level of products on consumer stock-out next term reactions. J Retailing 81(2):15–34

Smith SA (2009a). Clearance pricing in retail chains. In: Agrawal N, Smith SA (ed) Retail supply chain management. International Series in Operations Research & Management Science. Springer, Boston, MA

Smith SA (2009b). Optimizing retail assortments for diverse customer preferences. In: Agrawal N, Smith SA (ed) Retail supply chain management. International Series in Operations Research & Management Science. Springer, Boston, MA

Smith SA, Agrawal N (2000) Management of multi-item retail inventory systems with demand substitution. Oper Res 48:50–64

Stadtler H (2008) Supply chain management: An overview. In: Stadtler H, Kilger C (ed) Supply chain management and advanced planning. Springer, Berlin, pp. 1–36

Stadtler H, Kilger C (eds) (2008) Supply chain management and advanced planning: Concepts, models, software, and case studies. Springer, Berlin

Sternbeck M, Kuhn H (2010) Differenzierte Logistik durch ein segmentiertes Netzwerk im filialisierten Lebensmitteleinzelhandel. In: Schönberger R, Elbert R (ed) Dimensionen der Logistik. Gabler, Wiesbaden, pp. 1009–1038

Swami S, Eliashberg J, Weinberg CB (1999) SilverScreener: A modeling approach to movie screens management. Market Sci 18(3):352–372

Syring D (2003) Bestimmung effizienter Sortimente in der operativen Sortimentsplanung. Ph.D. thesis, Freie Universität, Berlin

Tellis GJ (1988) Price elasticity of selective demand: A meta-analysis of econometric models of sales. J Mark Res 25(4):331–341

Tellis GJ, Zufryden FS (1995) Tackling the retailer decision maze: Which brands to discount, how much, when and why? Market Sci 14(3):271–299

Tempelmeier H (2008) Material-Logistik: Material-Logistik: Modelle und Algorithmen für die Produktionsplanung und -steuerung in Advanced Planning-Systemen, 7th Edition. Springer, Berlin

Teo CP, Shu J (2004) Warehouse-retailer network design problem. Oper Res 52(3):396–408

Thonemann U, Behrenbeck K, Küpper J, Magnus K-H (2005) Supply Chain Excellence im Handel: Trends, Erfolgsfaktoren und Best-Practice-Beispiele. Gabler, Wiesbaden

Turley LW, Milliman RE (2000) Atmospheric effects on shopping behaviour: A review of the experimental evidence. J Bus Res 49:193–211

Urban TL (1998) An inventory-theoretic approach to product assortment and shelf-space allocation. J Retailing 74(1):15–35

Urban TL (2005) Inventory models with inventory-level-dependent demand: A comprehensive review and unifying theory: Decision-aid to improve organisational performance. Eur J Oper Res 162(3):792–804

Urbany JE, Dickson PR, Sawyer AG (2000) Insights into cross-and within-store price search: Retailer estimates vs. consumer self-reports. J Retailing 76(2):243–258

van der Vorst J, Tromp SO, van der Zee DJ (2009) Simulation modeling for food supply chain redesign: Integrated decision making on product quality, sustainability and logistics. Int J Prod Res 47(23):6611–6631

van Donselaar KH, Gaur V, van Woensel T, Broekmeulen RA, Fransoo JC (2010) Ordering behavior in retail stores and implications for automated replenishment. Manag Sci 56(5):766–784

van Donselaar KH, van Woensel T, Broekmeulen R, Fransoo J (2006) Inventory control for perishables in supermarkets. Int J Prod Econ 104(2):462–472

van Heerde HJ, Leeflang PSH, Wittink DR (2004) Decomposing the sales promotion bump with store data. Market Sci 23(3):317–334

van Nierop E, Fok D, Franses PH (2006) Interaction between shelf layout and marketing effectiveness and its impact on optimizing shelf arrangements: Working Paper. ERIM Report Series. Research in Management, 1–35

van Ryzin G, Mahajan S (1999) On the relationship between inventory costs and variety benefits in retail assortments. Manag Sci 45:1496–1509

van Woensel T, van Donselaar KH, Broekmeulen RACM, Fransoo JC (2007) Consumer responses to shelf out-of-stocks of perishable products. Int J Phys Distrib Logist Manag 37(9):704–718

van Zelst S, van Donselaar K, van Woensel T, Broekmeulen R, Fransoo J (2009) Logistics drivers for shelf stacking in grocery retail stores: Potential for efficiency improvement. Int J Prod Econ 121(2):620–632

Verbeke W, Farris P, Thurik R (1998) Consumer reponse to the preferred brand out-ofstock situation. Eur J Market 32(11/12):1008–1028

Viswanathan S, Mathur K (1997) Integrating routing and inventory decision in one warehouse multiretailer multiproduct distribution systems. Manag Sci 43:294–312

Waller MA, Heintz Tangari A, Williams BD (2008) Case pack quantity's effect on retail market share: An examination of the backroom logistics effect and the store-level fill rate effect. Int J Phys Distrib Logist Manag 38(6):436–451

Walter CK, Grabner JR (1975) Stockout cost models: Empirical tests in a simulation. J Market 39(July):56–68

Wang X, Liu L (2007) Coordination in a retailer-led supply chain through option contract. Int J Prod Econ 110:115–127

Whitin TM (1957) The theory of inventory management. Princeton University Press, NJ

Whybark DC, Yang S (1996) Positioning inventory in distribution systems. Int J Prod Econ 45:271–278

Wickern J (1966) Store space allocation and labor costs in theory and practice. J Retailing 42(Spring):36–43

Wong CY, McFarelane D (2007) Radio frequency identification data capture and its impact on shelf replenishment. Int J Logist 10(1):71–93

Xin G, Messinger PR, Li J (2009) Influence of soldout products on consumer choice. J Retailing 85(3):274–287

Yan XS, Robb DJ, Silver EA (2009) Inventory performance under pack size constraints and spatially-correlated demand. Int J Prod Econ 117(2):330–337

Yang M-H (2001) An efficient algorithm to allocate shelf space. Eur J Oper Res 131(1):107–118

Yang M-H, Chen W-C (1999) A study on shelf space allocation and management. Int J Prod Econ 60–61(4):309–317

Yin R, Aviv Y, Pazgal A, Tang CS (2009) Optimal markdown pricing: Implications of inventory display formats in the presence of strategic customers. Manag Sci 55(8):1391–1408

Yücel E, Karaesmen F, Salman FS, Türkay M (2009) Optimizing product assortment under customer-driven demand substitution. Eur J Oper Res 199(3):759–768

Yue X, Liu J (2006) Demand forecast sharing in a dual-channel supply chain. Eur J Oper Res 174:646–667

Zenor M (1994) The profit benefits of category management. J Market Res 31(May):202–213

Zhao QH, Wang SY, Lai KK, Xia GP (2004) Model and algorithm of an inventory problem with the consideration of transportation costs. Comput Ind Eng 46(2):389–397

Zinn W, Liu PC (2001) Consumer response to retail stockouts. J Bus Logist 22(1):49–71

Zolfaghari S, El-Bouri A, Namiranian B, Quan V (2007) Heuristics for large scale labour scheduling problems in retail sector. INFOR 45(3):111–122

Zoryk-Schalla AJ, Fransoo J, de Kok TG (2004) Modeling the planning process in advanced planning systems. Inform Manag 42(1):75–87

Zufryden FS (1986) A dynamic programming approach for product selection and supermarket shelf-space allocation. J Oper Res Soc 37(4):413–422